The Fly in the Cathedral

By the Same Author

Rain

Jill Dando: Her Life and Death

The Case of Stephen Lawrence

Were You Still Up for Portillo?

Test of Greatness: Britain's Struggle for the Atom Bomb

THE FLY
IN THE CATHEDRAL

———————

BRIAN CATHCART

FARRAR, STRAUS AND GIROUX
NEW YORK

Farrar, Straus and Giroux
19 Union Square West, New York 10003

Printed in the United States of America
Originally published in 2004 by the Penguin Group, Great Britain
Published in 2005 in the United States by Farrar, Straus and Giroux
First Farrar, Straus and Giroux paperback edition, 2006

The Library of Congress has cataloged the hardcover edition as follows:
Cathcart, Brian.
The fly in the cathedral / Brian Cathcart.— 1st American ed.
    p. cm.
Includes bibliographical references and index.
ISBN-13: 978-0-374-15716-6
ISBN-10: 0-374-15716-2 (hardcover : alk. paper)
1. Science—England—Cambridge—Biography. 2. Science—England—
Cambridge—History. 3. Rutherford, Ernest, 1871–1937. 4. Walton, Ernest,
1903–1995. 5. Nuclear fission. 6. Radioactivity. I. Title.

Q141.C2515 2004
539.7'62—dc22

                                                        2004056348

Paperback ISBN-13: 978-0-374-53026-6
Paperback ISBN-10: 0-374-53026-2

www.fsgbooks.com

1   3   5   7   9   10   8   6   4   2

Contents

List of Illustrations vii

Foreword xi

Prelude: Manchester, 1909 1

1. Cavendish 9

2. 'Mollycewels an' Atoms' 20

3. Method 36

4. A Way Forward 49

5. A Man in White Trousers 66

6. A Finite Probability 85

7. Hardware 101

8. Lab Life 112

9. Other Ideas 131

10. Turning Point 152

11. Off to the Races 176

12. Timeliness and Promise 201

13. Red Letter Day 223

14. Still Safe 244

15. Nobel 261

Postscript 272

Notes 275

Acknowledgements 290

Bibliography 293

Index 299

List of Illustrations

Inset

1. The Cavendish Laboratory, Free School Lane entrance. (Cavendish Collection)
2. Ernest Walton, Ernest Rutherford and John Cockcroft posing for news photographers outside the lab on the morning of 2 May 1932. (© Hulton-Deutsch Collection/CORBIS)
3. James Chadwick, assistant director of the Cavendish. (© Bettmann/CORBIS)
4. Cockcroft with George Gamow. (Cavendish Collection)
5. Winifred Wilson and Ernest Walton in the countryside outside Belfast. (Walton family collection)
6. Elizabeth and John Cockcroft at their home in Cambridge. (Cockcroft family collection)
7. The Cockcroft-Walton apparatus in 1930. (1851 Royal Commission Archive)
8. Merle Tuve. (Emilio Segrè Visual Archives, American Institute of Physics)
9. Robert van de Graaff with the first prototype of his generator in 1931. (Emilio Segrè Visual Archives, American Institute of Physics)
10. Stanley Livingston and Ernest Lawrence standing inside the giant electromagnet for their 27-inch cyclotron. (Emilio Segrè Visual Archives, American Institute of Physics)

11. The rudimentary 'control table' for the 800,000-volt
apparatus in the former Lecture Room D, with
Cockcroft at work, from the *Illustrated London News*,
11 June 1932. (Courtesy of the *Illustrated London News*
Picture Library)
12. The acceleration tube, with Walton monitoring results.
(Courtesy of the *Illustrated London News* Picture Library)
13. The complete Cockcroft–Walton accelerator, pictured
in the *Illustrated London News*, 11 June 1932. (Courtesy
of the *Illustrated London News* Picture Library)
14. The core of the apparatus with which Chadwick
discovered the neutron. (Cavendish Collection)
15. The Solvay Conference of 1933. (Cavendish Collection)

Papers, courtesy of the Churchill Archives Centre,
Churchill College, Cambridge
p. 109: Walton's bulb drawing, from the Walton notebooks,
courtesy of the Churchill Archives Centre, Churchill
College, Cambridge
p. 166: The opening page of the first Royal Society paper of
1930 by Cockcroft and Walton, from *Proceedings of the
Royal Society*, London A129, p. 477 (1930). Courtesy of
the Royal Society
p. 188: Walton's sketch of the rectifier tower, from the
Walton notebooks, courtesy of the Churchill Archives
Centre, Churchill College, Cambridge
p. 206: Walton's sketch of the acceleration tube, from the
Walton notebooks, courtesy of the Churchill Archives
Centre, Churchill College, Cambridge

p. 233: Walton's sketch of the target arrangement, from the Walton notebooks, courtesy of the Churchill Archives Centre, Churchill College, Cambridge

p. 247: Front page from *Reynolds's Illustrated News*, 1 May 1932, courtesy of Special Collections, J. B. Priestley Library, University of Bradford

Foreword

I was a schoolboy in Dublin when I first heard that there was a man at Trinity College who had 'split the atom'. At the time I do not suppose it would have meant much more to me if I had been told he had split an infinitive, but the phrase lingered and the notion came to tease me. Even young boys know that atoms are so small that millions fit on a pinhead, so what tool could you possibly find that would split one? I was a student at Trinity by the time I learned this feat had been achieved as long ago as 1932, when aeroplanes were still held together with wire, television was just a clever idea and many homes had yet to discover electricity. Even the most powerful microscope on the market in 1932 relied on glass lenses and though it could look pretty closely at a pinhead it was miles from picking out individual atoms. Splitting atoms in such a world seemed all the more improbable.

I did not study science, in fact I have never had a physics lesson in my life, so the mystery of how it was done began to unravel for me only after I became a reporter and had to read some nuclear history as background for a story. No one can explain why these fascinations arise, and given my unscientific background this one is unusually perverse, but that is when it all began. Rutherford and Curie, Bohr and Heisenberg, Fermi and Oppenheimer: they simply enchanted me, and there in the midst of them all were Cockcroft and Walton, who split the atom. (Walton was the Irishman I had heard about years before.) The more I read of their extraordinary feat the more I wanted to know, and so I progressed from popular accounts to specialist histories, from there to obscure articles and memoirs and finally to original letters and personal reminiscences. This book is the result.

It may seem impertinent for a non-scientist to tackle such a subject but I take my licence from Rutherford himself. He used to tell young physicists that if their research could not be explained in terms comprehensible to a barmaid it was probably not worth doing. Patronizing to barmaids as it may be, I have found this idea inspiring. What it says to me is that, setting aside the algebra and the jargon (which I hope I have done), the story of how the atom was split is one of human dreams and human endeavour that may be, and perhaps should be, enjoyed by everyone. It goes to the very essence of the universe and yet it involves people with ordinary lives and loves who worked with nothing more exotic than steel, wire, glass, oil, a few odd minerals and their own good sense. It is an adventure, and in its way a heroic one. Splitting the atom was like reaching the South Pole or the summit of Everest, one of the moments in history when we have stretched out to touch the limits of our known world, except that in this case the journey is contained within the laboratory and the human mind.

Brian Cathcart
London, 2003

Prelude: Manchester, 1909

In a small room in a university laboratory a nervous young man sat in darkness, waiting. Outside there was daylight, the watery sunshine of an English afternoon in early spring, but here the heavy black blinds were down and all was deepest gloom. As silent minutes passed, the effect the young man was hoping for gradually took place: his eyes were growing accustomed to the conditions. He began to make out the dark rectangle of the door, the black lines between the bare boards of the floor, the shadowy form of the workbench where he had set up his apparatus. Still he waited, for he knew from his training that the pupils of his eyes were not yet fully dilated, as they had to be before he could proceed. And as he sat he ran over in his mind the preparations he had made for the experiment he was about to carry out, ticking off a mental checklist until he was sure nothing had been missed. Then he rehearsed step by step what he must do next if he was to be confident his results would be sound. Only after he had been sitting there for twenty minutes did he rise to his feet.

Taking care not to look at the bulb, he switched on a very weak electric lamp pointing into the apparatus, then perched himself on a stool and leaned forward to the microscope. He knew he would have to hold this position for some time, so to make the job more comfortable he had covered the outer part of the eyepiece with cork. Resting the side of his face lightly upon this cushion and deliberately relaxing his body as much as he could, he closed his other eye and began to focus.

His field of vision was now filled by a single square millimetre of opaque glass, magnified fifty-fold and faintly illuminated by the lamp, and no sooner was he settled than he became aware he could see flashes of light on the glass – flashes so small and fleeting that if his eye had not been attuned to the dark he might not have made them out. They were like distant, momentary stars in the night sky, except with a hint of colour, a yellowy green. He watched in surprise as they exploded silently on every part of the square millimetre. They came in an irregular, stuttering flow: now one, now three together, and then another and another, then a pair, and so on. Forcing himself to concentrate, he counted them and found that on average there was rather more than one flash per second. This was not at all what he had expected.

The young man was an undergraduate rather than a seasoned researcher and he had been assigned this experiment by Ernest Rutherford, his professor – so short was the distance to the frontier of knowledge in atomic science that Rutherford was in the habit of giving original research tasks to his more promising final-year students. Yet this investigation was not one from which the professor had expected to learn very much; in fact he saw it as a simple matter of elimination. The student was aware of this, just as he was conscious that his proficiency in the laboratory was being tested. He wrote later: 'I knew enough to appreciate that, even though a negative effect might be expected, yet if I missed any positive result it would be an unforgivable sin.'[1] His preparations, therefore, had been unusually thorough and he had taken great pains to follow every laboratory procedure, including that twenty-minute wait in the dark, to the letter.

We can imagine his thoughts as he looked through the microscope and saw, not the occasional stray spark the professor had led him to expect, but a stream of sparks. One part of his mind must have thrilled at the notion that he had made a discovery, however peculiar it might be, but the greater part, the rational part, surely told him that he had it wrong, that somewhere in his apparatus he had fitted a piece incorrectly, or miscalculated a distance or strength, or failed to allow for some form of interference. Professor Rutherford, after all, was the foremost living authority in this field, a man who had only the previous year won the Nobel prize; he was unlikely to be mistaken about something so fundamental.

For a week, therefore, the student kept his counsel while he checked and re-checked every aspect of the experiment. Many times in those days he sat drumming his fingers in the darkness, waiting to begin another round of observations and fighting the temptation to clip a few minutes off the prescribed delay. But no matter how careful he was, it still happened: every time he put his eye to the microscope he saw those beautiful, surprising little flashes. Eventually, after trying every test and variation he could think of, he convinced himself that there could be no error and that this effect was really occurring. He set off excitedly to look for the professor and found him descending the stairs from his office. There, with a pride 'something like that of a cat delivering a choice mouse to his mistress', he announced his results.[2] Rutherford was thunderstruck. Receiving that news, he would say many years later, was 'quite the most incredible event that ever happened to me in my life'.[3]

The year was 1909, the place was Manchester and the young man was called Ernest Marsden. What he had been watching, what those little flashes announced, was an effect taking place at a scale smaller than atoms. Each spark represented a single sub-atomic event. This was not new in itself – such events had been observed before – but that those flashes should occur in the circumstances of Marsden's experiment was beyond all expectation. In fact it turned the received wisdom about the interior of the atom on its head.

Already in those years it was well known that the atom of matter was not the irreducible, solid ball first imagined as long ago as ancient Greece, for it could be shown that some atoms *did* reduce, in fact some atoms could spontaneously erupt like volcanoes and cast out small fragments, a process known as radioactivity. Physicists had also proven the existence inside atoms of particles they called 'electrons', which were about two thousand times lighter than the lightest atoms. Atoms, therefore, were not the smallest entities in nature but were themselves made of something, or some things; they must have an interior of some kind. The dominant theory in 1909 was that they resembled plum puddings: the sponge of the pudding was a spherical, permeable bulk of mysterious electrification and the electrons were the plums, speckled in large numbers through and around it. But while this image seemed to account best for the accumulated experimental findings to date it certainly did not accord with Marsden's observations. He had fired into atoms particles that were much more massive than electrons, and the flashes he saw told him that some of his projectiles were bouncing back. In the plum pudding model this was impossible, for there was nothing inside that was remotely big enough to cause such a thing, nothing big enough to bounce off. As Rutherford would put it, it was as if 'you fired a fifteen-inch shell at a piece of tissue paper and it came back and hit you'.[4]

Marsden's findings were soon verified and not long afterwards, though still aged only twenty and without even a bachelor's degree to his name, he had the privilege of reporting them to a meeting of the Royal Society in London. At that point, although he was to work under Rutherford for a few more years, his involvement in this story ends; it is his discovery and its consequences that concern us here. At first that discovery was greeted with a strange silence in the scientific world – so baffling were these results that for a year and more they were scarcely discussed. In one man's mind, however, they were nagging away like a tiresome, repetitive noise, distracting him from other thought. For a decade Rutherford had maintained a prodigious flow of revelations and publications but now his productivity fell away as he wrestled with the meaning of that one experiment. It took about eighteen months but in the end the answer came, so that one morning he was able to sweep into his laboratory and announce with delight that he 'knew what the atom looked like'.[5]

This vision he formally revealed at a public meeting of a Manchester scientific society where, providing a touch of bathos, the previous speaker had reported on an exotic snake found in an imported cargo crate. What Rutherford had to say was more momentous, indeed it was revolutionary. The atom, he declared, was not a pudding or an amorphous, electrified mass but instead was almost entirely composed of empty space. It had no solid boundary and its outer limits were defined only by the movement of its outermost electrons, while at its centre lay an object whose existence young Marsden had detected for the very first time: the atomic nucleus. This nucleus was not merely a new discovery, like another little island to be added to a map; it was an entity far beyond the experience or comprehension of science at that time, and it was of the utmost importance. Inside it, Rutherford knew, must be concentrated all but an insignificant fraction of the mass of the whole atom, and this must be packed to an extraordinary density. It was, in short, the heart of matter, and on an unimaginably small scale. This was to become known as the 'solar system' model, in which every atom had an internal sun, a massive, powerful central body, while around this sun, through a region otherwise unoccupied, orbited a number of electrons moving at fantastic speeds. The flashes in Marsden's microscope occurred when particles streaming on to a target of gold atoms occasionally rebounded off a nucleus back on to his glass screen, causing a spark. So small was the nucleus that only one particle in 8,000 would collide with it in this way, the other 7,999 traversing the emptiness unaffected.

It is hard to imagine a more fundamental physical insight, for Rutherford was redefining all the materials of the universe. Henceforth every physical thing would be different in the eyes, or at least in the informed imagination, of humankind. What is this book, the book in your hand, made of? The answer, overwhelmingly, is empty space. We may think of such objects as being solid but all the truly solid matter in this book, all the tiny nuclei and all the even tinier electrons, if they were concentrated on one spot, would barely fill the full stop at the end of this sentence. In relation to the atom, the nucleus was a mere speck in a cavernous void, or to use an analogy popular at one time, it was like a fly in a cathedral.[6]

The theory of the nuclear atom was Rutherford's greatest scientific achievement, even more important than the earlier work which had won him the Nobel prize, but he did not and could not stop there, for having once glimpsed the fly in the cathedral he ached to know more, to catch it, examine it, dissect it. The nuclei of the various chemical elements – zinc, sulphur, neon and the rest – were obviously made of something, but what was that something? Was it different in each case or was it the same materials, rearranged in different quantities? Perhaps, inside the nucleus, there was a single building-block that was fundamental to all matter, or perhaps there were several different kinds of block. Whatever the answer, it was the key to everything around us, be it solid, liquid or gas. Rutherford saw all of this very quickly; he knew that his scientific journey would not be complete until he had some idea of the structure and composition of the nucleus. For that it would be necessary to break it open and look inside, a process that would eventually become known, in one of those loose journalistic phrases, as 'splitting' or 'smashing' atoms.

He was to live another twenty-six years and he devoted those years above all else to the pursuit of that objective. Just as Marsden had, so generations of other researchers trained with the great man, supported his quest for a few years and then moved on. Rutherford too moved on, leaving Manchester to become director of the great Cavendish Laboratory in Cambridge; he was knighted and then ennobled; he became President of the Royal Society and a member of the Order of Merit; he was showered with honorary degrees and medals to go with his Nobel prize and he became one of the most famous scientists in the world, but for all this personal glory he was never satisfied and he never relented in his hunt for the fly in the cathedral.

To his great joy, though the wait was a long one, he lived to see the new surge of discovery he had dreamt of, playing a prominent and vital role. And hard-won though it was, that breakthrough ranks among the most astonishing and unexpected scientific achievements of the twentieth century. It involved apparatus of a nature and scale never seen in university laboratories before; it was built upon bold, almost perverse theoretical insights and at the same time depended on mundane technical advances; it flowed from years of frustration but ended in a competitive rush that pitted Rutherford's laboratory against American rivals adopting different approaches. When the climax came the discovery made headlines around the world. This book tells the story of how the atom was split, and to begin it we will jump forward a few years, to 1927.

1. Cavendish

For many years Cambridge railway station was not to be found in Cambridge at all, but in the countryside a mile or so out of town. The maps show the line from London closing in on the city and then at the last moment veering eastwards as if repelled by invisible forces within. And repulsion by invisible forces was more or less what happened, for when the first railway was approaching in the 1840s the Cambridge colleges were so fearful of its influence that – in much the same spirit that they secured a ban on Sunday rail traffic – they contrived to locate the station at what one historian has called 'an inconvenient distance'.[1] Several times in later years there were proposals for a more central terminus but they all came to nothing and eventually it was the town that moved, houses and businesses steadily creeping out along the road towards the station until the two were joined and the green fields pushed into the background. That inconvenient distance from the old town centre remains, however, and new arrivals at the station can still be dismayed to find their true destination some way off.

At 8.50 p.m. on Monday 17 October 1927, Ernest Thomas Sinton Walton was just such an arrival. He was twenty-four years old, of medium height with a wiry build, a high forehead, heavy spectacles and a suit of clothes which, while perfectly respectable, bore no trace of style. He was tired, having taken the overnight ferry from Ireland and then changed trains twice as he worked his way across England. The carriages had been crowded and at each staging-post he had had to oversee the unloading and loading of a heavy trunk containing such items as his toolbox, his essential textbooks and, most precious of all, the draft of his M.Sc. thesis. Now he alighted amid the gloom and steam of Cambridge station and once again extracted his trunk from the baggage car. Depositing it in the left-luggage office, he made his way out through the arched portico and there discovered the quirk of his location: he had not quite arrived. It was too late, in any case, to try to make contact with anyone in the town so he found a hotel nearby and had an early night.

After breakfast the next morning he set about his business. It was fortunate that he had an equable personality for another man in his position might have been anxious. There had been an unfortunate mix-up over his application to become a research student at the Cavendish Laboratory, with the result that while other successful applicants had arrived weeks earlier he had needed a last-minute scramble to secure his place. In fact there were grounds to suspect that the laboratory had accepted him only with reluctance, so a warm welcome was by no means assured. This was bad enough, but when Walton made it into town that morning a more pressing concern soon presented itself: time was ticking by but nowhere in the medieval maze of streets and passages could he find his destination and no passer-by whom he approached was able to guide him. 'Cambridge,' he wrote a couple of days later, 'is the hardest place I ever saw to find your way through. I must have spent over half an hour looking for the Cavendish Laboratory, and I scarcely know the way to it yet, there are so many turns and streets to go through.'[2]

Free School Lane is little more than an alleyway at the back of one of the older colleges, but half-way along it a relieved Walton at last came across the tall, grey Victorian building he was looking for. There was nothing to announce its identity unless you counted a statue of the Duke of Devonshire (family name: Cavendish) and an inscription in Latin which, when translated, announced: 'The works of the Lord are great, searched out by all who have delight in them.'[3] Walton made nothing of these clues but he had been assured this was the place and the oak doors were wide open, so in he went. An archway led beneath the body of the building towards a cobbled courtyard crowded with parked bicycles, and on the right was an office. Asking to see the director, he was informed that Sir Ernest Rutherford was away and that he should report instead to the assistant director. Stone stairs took him up to a dingy corridor where he soon found the office of Dr James Chadwick, a lean man of thirty-five with spectacles, an intense gaze and not much conversation. That Walton had failed to turn up at 9 a.m. sharp was of no consequence to the assistant director since the laboratory tended to begin the day in leisurely fashion. (Punctuality mattered much more when it came to going home.) In any case Chadwick had other things on his mind. Briskly but politely he did what he usually did with new researchers from overseas and sent the young man off to register as a member of the university and be assigned somewhere to live.

The University of Cambridge has many colleges, most of them of ancient foundation, and it was to Trinity that Walton made his way. This was the largest and the richest of them all, but it was also the college most closely associated with the Cavendish. Walton was duly accepted there by the senior tutor, who sent him on to the junior bursar, who after a short interview dispatched him to number 4 Park Parade, a few streets away at the north end of town, to view some lodgings that had just been vacated. Walton liked these. 'They are very clean,' he wrote to his father, 'and the sitting room, which I have to myself, is very comfortable. It is completely furnished and has two easy chairs and a small Chesterfield.'[4]

The bedroom, upstairs in a sort of attic, was less commodious and the bed was hard, but the rent, at £11 per term, was considerably cheaper than rooms in college, and within his budget. Breakfasts and the use of his gas fire, he noted, were not included in the price and the electric light was also an extra, at £1 per term. The rest of his day was filled with collecting his trunk from the station and opening an account at the Westminster Bank, and in the evening Walton had his first experience of dinner in college, a five-course affair beginning at what he considered the late hour of 8 p.m. It impressed him greatly and he could not help drawing the comparison with his *alma mater*, Trinity College, Dublin. 'There is a huge dining hall and the place is swarming with waiters, in fact the style leaves T.C.D. far behind. You can order almost anything you like either in the hall or to be sent round to your rooms . . . but of course it all appears on the bill.'[5]

The following morning he found his way once again to the laboratory and met Rutherford himself. He did not record his first impressions of the great man except to say that he seemed 'very nice', but an Australian student who arrived that same autumn, Mark Oliphant, has left a vivid picture of a young researcher's first encounter with the Cavendish professor.

When my turn came, I entered a small office littered with books and papers, the desk cluttered in a manner which I had been taught at school indicated an untidy and inefficient mind. It was raining, and drops of water ran reluctantly down the grime-covered glass of the uncurtained window. I was received genially by a large, rather florid man, with thinning fair hair and a large moustache, who reminded me forcibly of the keeper of the general store and post office in a little village in the hills behind Adelaide where I had spent part of my childhood. Rutherford made me feel welcome and at ease at once. He spluttered a little as he talked, from time to time holding a match to a pipe which produced smoke and ash like a volcano.[6]

Walton had sent ahead of him an account of the M.Sc. research he had done in Dublin and after the opening formalities the professor referred to this. The work was not in atomic physics but hydrodynamics and Rutherford said he had showed it to a colleague in that field who liked it a great deal. In particular he had praised the photographs Walton had been able to take of a curious effect in water, believing they were the clearest of their kind to date. This was the best sort of impression to make, for to Rutherford nothing was so important as to be able to carry off an experiment well. There seems to have been no hesitancy in the professor's welcome either, despite the problems over the application, and so when Walton bade Rutherford farewell and stepped out of that scruffy, smoke-filled office he was able to go straight upstairs and take his place in the laboratory's induction course. In this quiet way he joined what was a remarkable community.

There was nowhere in the world quite like the Cavendish. Founded in 1874 with funds from the Devonshire family, it was already, in 1927, a place of illustrious tradition. All four of its directors had been of international stature, physicists who remain to this day, if not quite household names, at least prominent in the histories and textbooks. Rutherford's immediate predecessor, Joseph John Thomson, or 'J. J.' as he was known, had discovered the electron. Thomson had succeeded Lord Rayleigh, an experimenter who broke new ground in the study of light and sound and was the discoverer of the noble gas, argon. And Rayleigh's predecessor, the first director, was James Clerk Maxwell, the great Scottish theorist of electromagnetism and a heroic figure in nineteenth-century science. The thrill of a heritage so rich, accumulated in barely half a century, must have impressed itself upon every young researcher entering the workrooms and lecture theatres of the laboratory, but the reputation of the Cavendish did not depend on history alone. Including Rutherford himself, in 1927 there were three Nobel prizewinners on the staff and a fourth researcher was to receive the prize that winter, while among the younger scientists a further four or five (including Chadwick) were already acknowledged as world leaders in their fields. Such concentration of talent was exceptional in a world where the age of gentlemen researchers working alone in private laboratories was fresh in many memories, but under Rutherford's leadership the laboratory in Free School Lane had grown accustomed to taking its own path.

There was the matter of numbers: by the standards of the time the Cavendish was a veritable factory for postgraduate physicists. Each year it admitted fifteen or more, roughly half of whom had taken their primary degrees at Cambridge. Of the rest a few came from other British universities but most were from overseas, principally Canada, Australia and New Zealand but also the United States, India, Japan, Italy, Russia and other European countries. At any given time, then, there were around forty people carrying out postgraduate research and in the context of the British university scene in the 1920s this was mass production. What the Cavendish expected of them was that they would do original work, suitably supervised and if possible published, and then leave with a Ph.D. if they were good enough. Some would need three years for this but others who had already done research at their previous university might be out in two. The objective in educational terms was straightforward, as Chadwick was to explain: 'We hoped that they would get the proper attitude to research

[and] acquire the ability to tackle any problem in physics.'[7] Thus equipped, they might go to other universities, to industry or government service or into schoolteaching, while one or two in each year who showed particular promise would stay on as researcher-lecturers if space and funds could be found. Besides this general goal of raising the national competence in science and spreading the word, the laboratory and its director had an additional objective, and this was another distinguishing mark of the Cavendish.

Rutherford once observed that all true science was physics and the rest was stamp collecting. Within the walls of his own laboratory he took this approach one step further, relegating all physics that did not concern itself with the atom to the philatelic category. Rutherford and Chadwick deliberately directed the work of the Cavendish towards the study of the interior of the atom, and so far as was practical towards the nucleus. Barring a few senior figures pursuing their own inquiries and a handful of cases among the students, all the researchers were therefore rowing in the same direction, to the same beat, and the result was a domination of the subject that amounted almost to monopoly.

Between the discovery of radioactivity in 1896 and the end of the First World War most of the experimental work on the atom had been done in three countries: Britain (with its dominions), Germany and France, with Britain making the largest contribution. The war and its aftermath severely disrupted Germany's efforts, and though research in the United States was slowly growing in scale and ambition there is no doubt that as the 1920s advanced British dominance increased markedly – and in Britain it was Cambridge that was making the running. Atomic physics was widely viewed as the most exciting field in all science and if you were a young researcher hoping to make a contribution, then no matter where in the world you came from your best course was to get yourself to the Cavendish.

So it had been with Ernest Walton. He was the son of a Methodist minister who, following the policy of the Church, moved frequently between parishes in Ireland. It had been natural, therefore, to send the boy off to boarding school in Belfast for his education and there he proved a star pupil, with a special talent and enthusiasm for science and mathematics and also a gift for making things. One family story has him reading geometry textbooks at home for pleasure, while on his birthdays he liked to be given tools to equip what grew into a sophisticated home workshop. A surviving school report describes him as 'swift, sure, penetrating and comprehensive as thinker and worker', and declares that he showed 'brilliant promise'.[8] Sure enough, when he found his way to university in Dublin he carried all before him in mathematics and physics and on graduating in 1926 (with first-class degrees in both subjects) he stayed on as a postgraduate, choosing physics for the practical reason that it allowed him to use his hands as well as his brain. His research supervisor set him the task of studying the flow of liquids past cylinders, a matter of importance in civil engineering and aeronautics. The effect of this flow was known – it created twin groups of vortices in the liquid, like chains of small whirlpools – but Walton's job was to measure and explain it. Using an elegant apparatus of his own creation he photographed the vortices at each stage in their progress and then described the rules governing their formation and behaviour. This project was still under way when he learned of a much grander career possibility: a scholarship that might enable him to work under Rutherford.

The Royal Commission for the Exhibition of 1851, which exists to this day, has responsibility for investing and disbursing the considerable profits of Prince Albert's great enterprise. In

keeping with the Prince's enthusiasms it likes to support young
scientists and engineers and in the age of the Empire it was
especially keen on students from the dominions – Rutherford
himself, a New Zealander, had first come to Britain as an
'1851 man' thirty years earlier. Now in 1927 Walton learned
there was a chance that one of that year's awards might go to
a student from the Irish Free State, and with the enthusiastic
support of his professor he promptly applied, stating his wish
to study atomic physics at Cambridge. This was where the
muddle began. The Commissioners liked the application but
were unhappy about Walton's plan. He had made a promising
start in hydrodynamics, they wrote back; would he not do
better to continue in that line of research? It was a reasonable
question, though it may also betray a resentment that existed
in scientific circles at the way Rutherford monopolized young
talent. Walton stuck to his guns, replying that he was keen on
atomic physics and it was an area where research was 'likely to
lead to important results'.[9] The exchange delayed the approval
process by three weeks and then a further month was lost as
the Commissioners' letter granting the scholarship went astray.
When the good news finally reached him Walton discovered
to his horror that, contrary to his expectation, the Commission
had taken no steps to establish that the Cavendish would
actually take him on – that was something he should have
done himself, at the outset.

A swift appeal to the Cavendish produced only the gloomy
reply from Rutherford that it was late in the day and 'there
has been a great rush of men to the Laboratory this year'.[10] A
number had already arrived. The professor did not close the
door completely, however, and invited Walton to spell out
his interests more fully. While he was doing that the 1851
Commissioners, apparently feeling a share of responsibility for
the confusion, stepped in to ask Rutherford if something

could be done for the young man, even at this late stage. Only days remained before the closing of university admissions when the professor finally gave his approval. His letter offering the place must have reached Dublin after 11 October and it was just six days later that Walton turned up at Cambridge station.

For all his troublesome determination to study it, the Irishman was little better than a beginner in atomic physics. Trinity College, Dublin in his day was a scientific backwater and the undergraduate course he had taken was, by his own subsequent estimation, ten or fifteen years behind the times, so that whatever he already knew about the atom he must have picked up largely by his own effort. Most others in that 1927 Cavendish intake were better prepared. The Cambridge graduates among them already had two or three years of exposure to Rutherford's laboratory as undergraduates and so were perfectly placed, while several of the overseas arrivals had received a more up-to-date training at their own universities, often taught by Cavendish products from previous years. All researchers, however, were required to spend at least a little time in what was called the Nursery, the course for newcomers taught by G. F. C. Searle, an elderly physicist whose Father Christmas appearance belied a stern nature. From the middle of the summer onwards, Searle's Nursery functioned in a cramped room high up under the Cavendish eaves where benches were always crowded and it was always too hot or too cold.

Here Walton learned the techniques used in the laboratory and at the same time encountered some of the year's other new arrivals. Among them was that Australian, Mark Oliphant, a big, busy, friendly man with a shock of vertical brown hair, who was almost blind in one eye. At twenty-six he was two years older than Walton and, unusually for a new research student, married. He had taken longer to reach this stage in his career because he had had to pay his own way through university in Adelaide, which he did by working as a lab technician. Now he was in a hurry. Another of the 1927 intake was Norman de Bruyne, a bouncy, schoolboyish figure whose heavy brows and big spectacles could not conceal the sharp eyes and quick mind behind them. Though born to a Dutch father in southern Chile, De Bruyne was very English and had been to school at Lancing with the likes of Evelyn Waugh and Tom Driberg. As one of the Cambridge graduates in the intake he already knew his way around the Cavendish but he had also built up useful experience elsewhere, doing vacation work in a power company laboratory in London. A third to start that autumn was a Canadian, George Laurence, originally from Prince Edward Island and a graduate of Dalhousie University in Nova Scotia. Dalhousie had a record of research in radioactivity and Laurence already had some original work to his name; like Oliphant he was eager to get started on a project of his own. In such company – there were others, from Britain and elsewhere, and they were no less impressive – Ernest Walton passed six or seven weeks in that lofty attic room, building and testing standard apparatus, mastering vacuum equipment, gaining experience with radioactive materials and, no doubt, struggling to keep warm in the encroaching Cambridge winter.

2. 'Mollycewels an' Atoms'

When Rutherford was a young research student at the Cavendish in the closing years of the nineteenth century he worked alongside a French physicist, Paul Langevin. Long afterwards Langevin was asked to look back on his relationship with the rising star of experimental physics: had he and Rutherford quickly become friends? The reply is often quoted: 'One can hardly speak of being friendly with a force of nature.'[1] A force of nature was what Rutherford remained all his life, a barrelling, thundering, penetrating presence in the world of physics, a great rowdy boy full of ideas and energy who was guaranteed to challenge assumptions and conquer hearts and minds. Another good witness is James Chadwick, who worked with him for more than twenty years and probably knew him better than any other scientist. Asked once whether Rutherford had an 'acute mind', Chadwick decided that 'acute' was the wrong word. 'His mind was like the bow of a battleship. There was so much weight behind it, it had no need to be sharp as a razor.'[2] All who worked with him felt this; his thirst for knowledge, his determination to make the next discovery and his sheer intellectual momentum were so great that anyone standing close was likely to be swept along. A classics professor thrilled by a Rutherford lecture captured the spirit: 'Here was that rarest and most refreshing spectacle – the pure ardour of the chase, a man quite possessed by a noble work and altogether happy in it.'[3]

Born into a country family of modest means in New Zealand's South Island, Rutherford shone as a student from early schooldays and when he reached Cambridge made an instant mark with pioneering work on radio waves. A description survives of an occasion when he demonstrated this near-magical phenomenon to friends in his lodgings: at the appointed moment a signal transmitted over the ether from apparatus half a mile away rang a bell on his living-room table, prompting rapturous applause from the assembled students. But on the discovery of radioactivity in 1896 he abruptly dropped this work and switched to the new field, scoring a string of successes so swift and dramatic that within two years, at the age of twenty-seven, he was a professor in Montreal. There the pace of his discoveries actually quickened and a pattern was set as he drew in a succession of other scientists as collaborators, sometimes from different disciplines and often at the cost of their previous studies. Many found it the most exciting period of their whole careers. From Canada Rutherford moved in 1907 to Manchester, where over the next twelve years – though his research was interrupted by a spell of war work – he consolidated his standing as the foremost experimental physicist of the age. This is not the place to list his achievements but it is fair to say that of all the important discoveries and insights about the atom in these years Rutherford had a hand in at least half, and of these his concept of the nuclear atom is the crowning glory. So in 1919, when J. J. Thomson stepped down as director of the Cavendish in Cambridge, he was the obvious successor.

For all his prodigious ability, however, he was a man of lurid inconsistencies. Kind almost to a fault, he was always concerned about the lives and careers of his students and would find positive things to say about them in job references even when others struggled. With his peers he was no less generous – when Marie Curie became engulfed in scandal, for example (she had an affair with the married Langevin), he went out of his way to show support. It was said that few men ever made more friends and fewer enemies, and yet this was not something he achieved by being an easy companion or colleague, for he was anything but. His rages were ferocious, unpredictable and indiscriminate. If something went wrong in the lab or if a researcher did something of which he disapproved he could rail and sulk for hours and even days, venting his wrath not only on the culprit but on anyone, high or low, who crossed his path. In these moments his criticisms, delivered at maximum volume from close range, could be withering. The 'boys' in the workroom next to his office came to regard it as their inescapable fate, as the closest victims to hand, to be roasted by the professor every few weeks. Once the storm passed, however, Rutherford was usually contrite, apologizing sheepishly and making some quiet gesture to repair the damage. Everyone at the Cavendish had to live with this.

Something of the same tension was evident in his laboratory work. As an experimenter he was ingenious in the extreme, but no one could call him dextrous. 'Don't shake the bloody bench, Crowe,' he would bellow at his assistant when he himself was causing some difficulty with the apparatus.[4] Crowe's predecessor remembered Rutherford as 'horribly impatient', with unsteady hands and a habit of bumping into things – 'Give him a pair of tweezers and he would drop them and become cross.'[5] But this clumsiness appeared to have strict limits, for just as Rutherford never allowed his tantrums to cost him a friend so it was rare for him actually to damage his apparatus. The technician recalled: 'I don't know whether he had some reserve power that told him the elasticity of glass . . . but he never broke it.'[6]

A big man with a loud voice and – as many who met him remarked – the air of a colonial farmer, he was not, however, the sort to clap another on the back in a moment of shared pleasure or to place a hand on a shoulder as a gesture of sympathy. So far as he could he avoided touching others and his handshake was limp and reluctant. The diffidence extended to his home life. He was married to a woman he had known since New Zealand days and they lived in some comfort in Newnham Cottage, a pretty house a short walk from the lab on the other side of the Cam. They had a daughter and grandchildren but in front of others the couple never showed a hint of physical affection – rather the contrary, since the years made Mary Rutherford shrewish. A teetotaller, she would harry her husband mercilessly if he drank so much as a glass of sherry, while at table in front of guests she did not think twice about barking: 'Ern, you're dribbling again!'[7]

Rutherford's background had not been marked by warmth. In his childhood his mother Martha had been a domineering presence and in some respects this long-lived woman over-shadowed the whole of Ernest's life. Even in middle age he wrote her dutiful letters every fortnight or so, wherever he was, and though they are now lost it is clear the son struggled to please a mother who was distant in more ways than one. A poignant and discomforting peak was reached in 1931, in the telegram he sent informing her of his elevation to the peerage. 'Now Lord Rutherford,' he wrote. 'Honour more yours than mine.'[8]

He was generally a buoyant figure none the less, full of laughter and home-baked wit. He got about a great deal and met many people of importance, whether at high table in Trinity College, at international conferences or at official committees in London, and he relished the stories and repartee these encounters generated. No sooner was he back in the laboratory among his own people than he would spill these out, no doubt with embellishments, and the better ones were filed away for endless retelling. The pompous and ineffectual were favourite targets. 'That man,' he declared of some committee member, 'is like the Euclidean point. He has position but no magnitude.'[9] He loved the story of how he found himself alone in a college room once with a cleric, both of them waiting for different appointments. Breaking an awkward silence, he announced: 'Hello, I'm Lord Rutherford.' To which the cleric replied: 'Hello, I'm the Archbishop of Canterbury,' before silence fell again. The punchline – 'And I don't suppose either of us believed the other!' – would be delivered with a roar of delighted laughter.[10]

Beneath it all Rutherford was a perfectly tuned machine for scientific discovery. His energy, his memory, his concentration, his gift for visualization and his willingness to give everything a try took him to the forefront of his subject while his impatience, restlessness and unrelenting curiosity kept him there. He did not care for elegant apparatus; any scruffy lash-up would satisfy him so long as it did the job he wanted. His mathematics could appear so slapdash that even undergraduates were shocked – he seemed to switch pluses for minuses at his own convenience – but his results were almost always accurate enough for his purposes. The finer points he left to others – he loved to mock the sort of scientist who took pride in pushing a measurement one more decimal place – for he was always in a hurry to get on to the next thing.

In late 1927, when he began considering research assignments for the generation of graduate students that included Walton and Oliphant, Rutherford was fifty-six years old and well advanced in what was the third phase of his career: first Canada, then Manchester and now Cambridge. This Cambridge period was proving different from its predecessors in important ways. He was running a large laboratory and although a good deal of the administrative burden was carried

by Chadwick the responsibilities still occupied a greater proportion of his time and energy than before. He read theses
and conducted vivas, he vetted papers and decided whether
and where they should be published, he chose which projects
would go ahead and which would not. Above all, he determined the laboratory's policies and carried the responsibility
for them. And as if this was not distraction enough he was
now Sir Ernest, Nobel laureate and public figure, a personality
in demand in many quarters as lecturer, adviser, figurehead
and decision-maker. Public-spirited by instinct, he took a
conscientious view of these duties and they increasingly
removed him not only from the workbench but also from the
laboratory and even from Cambridge.

For a distinguished scientist in his fifties this progression
from the coalface of experiment to the committee room and
speaker's podium may be normal, but for Rutherford, the
battleship of physics, it was not. However pressed he was for
time he had not given up experimental work and had no
intention of doing so. Nor had he lost touch with his subject,
for he still read the important journals and articles as they
appeared, usually late at night in his study at home. Above all,
he had not lost his curiosity. No less than in 1911 when he
first described it, he was determined to get to grips with the
nucleus, still hungering for a close look at the heart of matter.
But there was the rub, for in this respect his Cavendish period
was proving a disappointment, the first of any significance in
his career. If we apply his own high standards we can say that
he had not made a really important discovery since 1919,
when he was still at Manchester. Nor, for that matter, had his
subordinates. It was true that one of the senior researchers,
F. W. Aston, had won a Nobel prize in 1922, and that another,
C. T. R. Wilson, was about to win the 1927 prize, but in both
cases it was for work done before Rutherford's arrival at the

Cavendish. The most successful research he had overseen in Cambridge in the atomic field was essentially the consolidation of his own 1919 experiments, carried out largely with Chadwick but also by a very bright former naval officer called Patrick Blackett. This had been gratifying, but it was not the pace to which Rutherford was accustomed.

One of Chadwick's minor duties was to translate any papers in the German scientific literature that referred to the director's researches, a job he could do off the cuff. Rutherford was vain enough to enjoy hearing himself praised in these papers and was especially fond of one German word, *bahnbrechend*, which means 'ground-breaking'. Whenever that came up he would have Chadwick read the sentence over again while he purred to himself in gratification. Now, as the 1920s tipped towards their close, Rutherford was conscious that there were fewer occasions for the word. The knowledge frustrated and annoyed him; it depressed him; in Chadwick's view it was the underlying cause of some nagging illnesses he suffered. It was not that his nuclear theory had been overturned; on the contrary, the idea of the fly in the cathedral was by now universally accepted. It was the internal workings of the fly and the rules that governed its behaviour that were proving elusive. So what was known, what had been discovered, and what were the remaining mysteries?

'Scientifically speakin', it's all a question of the accidental gatherin' together of mollycewels an' atoms,' declares a philosophical Dubliner in Sean O'Casey's play *The Plough and the Stars*. He goes on:

Mollycewels is a stickin' together of millions of atoms o' sodium, carbon, potassium o' iodide, etcetera, that, accordin' to the way they're mixed, make a flower, a fish, a star that you see shinin' in the sky, or a man with a big brain like me, or a man with a little brain like you!¹¹

It is a fair account of the facts and those facts were well known in 1926 when the play was first performed. Molecules do indeed make up the flowers, the fishes and the human brain and molecules are, as the clever fellow says, combinations of the atoms of the basic materials of our world: not just sodium, carbon and potassium but oxygen, copper, sulphur, magnesium, gold, phosphorus, mercury and others – the elements. In the 1920s it was thought there was a total of ninety-two of these elements and the atoms of each were unique to it – a tin atom could only be a tin atom; an atom of calcium could only be an atom of calcium and so on. These atoms differed from one another in a variety of ways of which the most obvious was their weights, and it was one of the great achievements of science up to 1927 to have established what those weights were.

Of course no one had placed individual atoms on weighing scales – they were far too small for that – but the chemists and physicists of the nineteenth century had found that they could deduce the relative weights of elements from their behaviour under various kinds of experiment. For example, they weighed a certain volume of water, split the water into its component elements, oxygen and hydrogen, and then measured the volume and weight of each. From this, and knowing what they did about the composition of gases, they were able to establish that a given large number of oxygen atoms weighed sixteen times more than a similar number of hydrogen atoms. It followed that a single oxygen atom must be sixteen times heavier than a single hydrogen atom. Similar studies of other elements supplied a whole list of weight relationships like this and in time a complete picture emerged. Two facts stood out:

first, that hydrogen was the lightest element, and second, that the weights of many of the other elements were approximately whole-number multiples of the weight of hydrogen. If you took hydrogen as 1, helium had a weight of 4, lithium 7, beryllium 9, boron 11, oxygen 16 and so on, the scale continuing up past silicon at 28, titanium 48 and platinum 195 to the heaviest element of them all, uranium, at 238. These figures were called the 'atomic weights'.[12] It was not a perfectly tidy table (for reasons not yet understood some elements did not emerge as whole numbers) but for the moment it was a useful tool.

Particularly useful, in fact, for physicists such as Rutherford who were interested in the make-up of the atomic nucleus. For a start they could see that these atomic weights were effectively the weights of the nuclei – the electrons in atoms weighed next to nothing and the rest was just empty space so the bulk of weight had to be in the nucleus. And if the weights of the various nuclei tended to be multiples of the weight of the hydrogen nucleus that seemed to suggest something very interesting: that nuclei were built of bricks of a standard size. There would be one of these bricks in a hydrogen nucleus, sixteen in an oxygen nucleus, forty-eight in a titanium nucleus and so on. All nuclei would therefore be clusters of the same sort of brick, but in different numbers. This idea soon won general approval and the standard brick was given a special name: 'proton'. The hypothesis appeared to be confirmed experimentally in 1919 when, by a process he called 'artificial disintegration', Rutherford managed for the first time to chip parts off nuclei. The part that came away was identical to a hydrogen nucleus – in other words it was a proton.

So now physicists knew of two fundamental sub-atomic particles, the electron and the proton. They were an odd couple. The electron was so slight a thing as to be almost nebulous and yet for something so small it had fantastic energy. It was, in effect, a hyperactive Tinkerbell whirling around inside the atom. The new proton was altogether quieter and more sober; its most striking characteristics were its mass and density – in fact it was reckoned to be 1,860 times heavier than the electron. Put one of each together and they formed a whole atom of hydrogen, with the proton as the nucleus or sun and the electron in orbit around it, a solitary planet. This much, the scientists could see, was straightforward. As a next step it might have seemed natural to expect that the atoms of all the other elements were variants of this model, with oxygen, for example, having sixteen protons in its nucleus and sixteen electrons spinning around it. But things were not so tidy.

Look at the case of helium, the next lightest and therefore next simplest element after hydrogen: with a weight of 4 it seemed that its nucleus must contain four protons, while it was known for certain that each helium atom held two orbiting electrons. Four and two: the numbers did not match. In uranium there was also a difference and it was far bigger: a weight of 238 implied that the uranium nucleus contained that number of protons, but it was known there were only ninety-two orbiting electrons. And there were also mismatches in all the other atoms in between. Why should this matter? Was there any reason why an atom should not contain more protons than electrons? There was, and it was a fundamental reason: the laws of electricity – an important force in atomic workings – simply forbade it. Electricity comes in two varieties, positive and negative, and the two types of sub-atomic particle reside in the different camps, the proton being positively charged and the electron negatively charged. Despite the difference in their weights, the *amount* of charge they carry is identical, so that one electron (−1) neatly cancels out one proton (+1). Now it was an iron rule of physics and a readily observable fact that every atom, taken on its own and as a whole, was electrically balanced and neutral, so it seemed to follow that every atom must contain equal numbers of protons and electrons.

Here was the great mystery confronting Rutherford and his colleagues. The iron rule was not to be questioned; the electrical charges in a whole atom, when taken together, were undoubtedly in balance. Yet the available evidence said quite the contrary: that inside all the atoms but hydrogen there were more protons than electrons. Something had to give.

A simple solution was to say that there had to be more electrons in the atom than the existing ideas envisaged – electrons were extremely light, after all, so adding a few more would not interfere with the known atomic weights. And there was only one place those extra electrons could be hiding: inside the nucleus. On this basis a helium atom would contain not only the two known, orbiting electrons but also two concealed ones, buried inside the nucleus. Two and two makes four: set these four (−4) against the four known protons in the nucleus (+4) and the whole atom would be electrically neutral as required. With heavier elements the sums would be bigger but the principle of the hidden 'nuclear' electrons remained the same: there would be enough to balance the books. It was a neat solution in some respects and it was not a purely expedient one, for there was a small amount of experimental evidence to support the idea of electrons existing inside the nucleus, evidence from the study of radioactivity. In the process of radioactive decay, as we have seen, the nuclei of a few heavy elements such as radium and uranium had the habit of ejecting some of their contents, volcano-like, and among the fragments that emerged, beyond all doubt, were electrons. If electrons were coming out of nuclei it seemed reasonable to conclude that they must previously have been inside them – in other words, electrons could indeed be components of the nucleus.

It is hardly surprising, therefore, that in the 1920s prevailing opinion held that the nuclei of all matter were probably made up of combinations of protons and electrons arranged in this way, and yet it would be wrong to infer that scientists were content with that picture. For a start they were uncomfortably aware that it was little better than guesswork, that if the components added up neatly in the new arithmetic it was only because they, the physicists, were making up the numbers to fit. Nobody actually knew that there were two electrons inside the helium nucleus, in fact nobody could prove that there were any electrons in there at all. Worse, there were practical problems about nuclear electrons. With protons alone as its components the nucleus already appeared to be improbably crowded, but no one could imagine how electrons might fit in there. It was not so much their size as their energy; those fizzing Tinkerbells would be highly disruptive. And more than that, everybody already knew what happened when a proton met an electron: they combined to become a hydrogen atom. Yet there was no room for hydrogen atoms inside nuclei.

When Rutherford had looked at this problem in 1920 he had come up with a theory of his own. The answer as he saw it was not that there had to be more electrons tucked away inside atoms but that there must be fewer protons in there. The books still balanced in his scheme, but instead of having four electrons and four protons in his helium atom he suggested two and two. The objection was obvious: helium had an atomic weight of 4, so Rutherford's arrangement would leave the nucleus too light. How could he answer that? He did so by proposing the existence of an entirely new particle, the neutron, which weighed the same as a proton but carried no electrical charge – it was a dumb entity, a makeweight. Under his alternative arrangement, therefore, the helium nucleus would contain two positive protons and two neutral neutrons and the atom as a whole would be kept in balance by its two orbiting electrons. The electrical characteristics of the nucleus and its weight were thus reconciled.

As Rutherford well knew, this was even more speculative than the notion of electrons inside nuclei. Protons and electrons existed and could be observed and studied but the neutron was nothing more than one man's fancy committed to paper. (In fact it was even more far-fetched than appeared at first glance because Rutherford believed that neutrons were really intimate combinations of protons and electrons, fused together in some extremely powerful way that he could not begin to explain.) Yet for various reasons he strongly preferred his theory and it provided the focus of most of his nuclear research in the 1920s: he was determined to find the neutral particle inside the nucleus and prove its existence experimentally. Working sometimes with Chadwick and sometimes alone, he tried on and off for years. By its very nature, however, the neutron was bound to be elusive because, having no charge, it would be unaffected by any attempts to tease it out by electrical means, while in terms of weight it would be impossible to tell it apart from a proton. If it did exist, in other words, it had powerful natural camouflage. Over the years the two men attempted every trick they could think of to lure it into the open and in time they became almost desperate, as Chadwick later described:

I did a lot of experiments about which I never said anything. Some of them were quite stupid. I suppose I got that habit or impulse . . . from Rutherford. He would do some damn silly experiments at times, and we did some together. They were really damned silly. But he never hesitated . . . There was always just the possibility of something turning up, and one shouldn't neglect doing say a few hours' work or even a few days' work to make quite sure.[13]

But nothing did turn up and this was the principal cause of the gloomy mood of the late 1920s. To use Chadwick's word, they felt they were in a morass. In the absence of proof of the neutron's existence, in fact, the whole idea languished to the point where scientists outside the Cavendish, who may have paid attention when it was first proposed, simply forgot about it. By contrast, the theory of electrons in the nucleus, unproven and unsatisfactory as it was, came to hold a clear ascendancy. Even Rutherford accepted it in public as the best available guess. Although privately he could not abandon his own idea, he knew it was senseless to continue pressing a case for which, despite sustained effort, no evidence could be found.

As if all this were not enough to keep the physicists occupied, there was another particle which seemed to exist inside the nuclei of some atoms and which they were quite unable to account for: the alpha particle. A hard, substantial object known to experimenters for years, and a particular favourite of Rutherford's, the alpha particle was not fundamental like the electron and proton but neither was it a ghost like the neutron. It was a composite, a bundle of particles with a weight of 4 and a charge of +2 – in fact it was identical to the nucleus of the helium atom. For some reason these alpha particles were abundant even in places where helium should not be, notably in experiments in radioactivity, where they were the biggest of the fragments cast out by radium and uranium. On the strongest available theory (the one which said that there were electrons inside nuclei) they were assumed to be made up of four protons and two electrons and it seemed that they too must exist as discrete entities within the nuclei of some more complex atoms. For some unknown reason, it was suggested, this arrangement was especially comfortable for nuclear particles, so that whenever the necessary parts existed inside a nucleus they would coalesce in that way.

Such, in outline, was the state of knowledge in the middle and later 1920s. It was certain that the nucleus contained protons, though no one had counted them, and there seemed to be no alternative but to accept that there were electrons in there too, though how they might fit was impossible to explain. The neutron was little more than a failed hypothesis, even if Rutherford and Chadwick had yet to give up on it. And then there were alpha particles, nuclei within nuclei found in some elements. Even to non-physicists it is clear that to call this a state of knowledge is an exaggeration; it was a state of confusion, if not of ignorance. And these were the easiest, most basic problems of the nucleus – beyond them lay, for example, the profound question of how it was all stuck together when logic said it should fly apart. To Rutherford it must have been an especially depressing picture. Fifteen or more years after he announced the existence of the nucleus it remained, to borrow from Churchill, a riddle, wrapped in a mystery, inside an enigma. Another physicist summed things up:

Certain generalisations are possible, certain tentative suggestions have been made, which seem helpful. On the other hand, the subject has offered a vast field for what the Germans call *Arithmetische Spielereien* [arithmetical games], which serve rather to entertain the players than to advance knowledge.[14]

Even in the Cavendish people were reduced to such game-playing. One young undergraduate had the temerity to present

Rutherford with a list of nuclear possibilities he had dreamt up in his spare time, in the belief he was being helpful. It can only have served to remind the professor how far the question closest to his heart had drifted into the world of idle speculation. This was, as Chadwick put it, 'in many ways a frustrating time'.[15]

3. Method

The principal cause of the frustration is not in doubt, at least in retrospect: this was a case where the workman could legitimately blame his tools. When we want to know how something works we usually take it apart. We unclip, unwind and unscrew until we can spread out the components and examine each in turn, then we fit them together in pairs, assemble larger combinations and finally reconstruct the whole affair. Such methods have been useful not only in technology but also in anatomy, and they had proved successful too in the physical sciences. Molecules were dismantled by exploiting the great range of natural chemical reactions to separate the elements and reveal them in isolation. And the atoms of those elements in their turn were teased apart mainly by the tricks of electricity, detaching electrons from their nuclei. With the atomic nucleus, however, the job of dissection was proving beyond the capability of the laboratory equipment of 1927.

A vivid light is shed on this problem by an academic dispute that was rumbling on at the time between the Cavendish Laboratory and two atomic researchers in Vienna, Hans Pettersson and Gerhard Kirsch. The argument had animated the journals, provoked angry exchanges at conferences and raised eyebrows among physicists the world over, and in the autumn of 1927 it was approaching its climax. The stakes were high. The Vienna pair were challenging the credibility of important Cambridge experiments, some of them carried out by Rutherford and Chadwick, and even threatening to shift the leadership in this field away from Britain. At the heart of the disagreement was the principal experimental technique then in use to probe the nucleus.

This technique was little different from that employed so successfully by Marsden in Manchester back in 1909. The apparatus which the young man had built then, and which enabled him to see the unexpected sparks that ushered in the nuclear theory, was a disarmingly simple one. It employed a tiny quantity of radioactive radium held in a container with a narrow aperture. In radioactivity atoms expel three kinds of fragment: the relatively large alpha particles, the much smaller beta particles (which are electrons) and the even more amorphous gamma rays (which are like light or X-rays). Even the smallest laboratory quantity of radium comprised billions of atoms so that alpha, beta and gamma rays would constantly radiate out from it in all directions. Marsden's container was designed to stop most of these rays while allowing a single, slender beam to emerge through the aperture. This continuous beam provided ammunition; what was needed next was a target. Following procedures pioneered by Rutherford, therefore, Marsden aimed the beam at a thin gold foil. All experience, consistent with the idea of the plum-pudding atom which then prevailed, suggested that every particle in the beam would pass clean through the gold just as bullets from a machine gun would pass through the spongy part of a pudding. The young man's task was to test that hypothesis, but how was he to tell what was happening?

By an almost miraculous quirk of nature there was a tool which allowed the researcher to see, if not quite the sub-atomic particles themselves, then the footprints they left behind. This was the scintillation screen, a device serendipitously discovered by one of the last of the great gentlemen scientists, Sir William Crookes. Crookes was completing an experiment in his private laboratory one day in 1903 when he managed to lose sight of the trace of radium he had been using. He was not concerned, for he knew that radioactivity excited a glow in a cheap compound called zinc sulphide, so he took some of this and moved it around the surface of his desk until, coming near to the radium speck, it glowed. In that moment, for some happy reason, Crookes began to think about the glow rather than the radium and he got out his magnifying glass to take a closer look. To his surprise he found that the light emerging from the chemical was not uniform but sparkly, like a series of tiny, individual events, fascinating and beautiful to behold. (So beautiful, in fact, that in a short time an adult toy was on sale in London which produced the effect in a sort of radioactive kaleidoscope. It was called a 'spinthariscope'.) More important than its visual charm, however, was what was causing the effect. There was sometimes a dull background glow, which Crookes correctly deduced was caused by beta and gamma rays, the lighter kinds of radioactive emission, but the sparks or 'scintillations' were something quite different. Each was the result of the impact on the zinc sulphide of a single alpha particle. It was an oddity no one could explain but it was to prove a very helpful one. Because of their size and speed alpha particles were the most useful radioactive ammunition for experimental purposes and Crookes had discovered a means not only of detecting their presence but also, with the help of a good microscope, of counting them.

When Marsden set up his apparatus it was thus natural for him to incorporate a glass screen dusted with zinc sulphide powder. This he placed – and here was the novelty of his experiment – on *the same side* of the gold foil as his radiation source, angled in such a way that alpha particles rebounding off the gold target would strike it. Because the particles were all expected to go straight through the gold it was generally assumed that he would see nothing, what he called a 'negative result'. Instead, as we know, he got a positive result and was able to watch many scintillations on his screen. From this outcome, and after carefully counting the flashes to establish what proportion of alpha particles rebounded, Rutherford was able to deduce the existence and approximate size of the object provoking the rebounds – the atomic nucleus.

There were thus three essential components to such an experiment: the radiation source, the target and the scintillation screen. This combination became the stock in trade of Rutherford's laboratory and it enabled him ten years later to announce another of his *bahnbrechend* discoveries. Working during breaks from wartime research on methods of detecting submarines, he took a beam of alpha particles and fired it through a cylinder of nitrogen gas. The result was a reaction which again caused the scintillation screen to light up, but this time there was something different about the sparks. Though still quite distinct, they were undeniably smaller than the familiar alpha particle impacts. What could they be? Rutherford solved the problem with his customary bravura. The new, smaller sparks were caused when *protons* struck the screen, and since these protons were not present in the original radioactive beam there was only one place they could have come from: the nitrogen. This was artificial disintegration and it confirmed for the first time the presence of protons in the nucleus. What was happening was that some of Rutherford's alpha particle projectiles were striking the nuclei of nitrogen atoms in the cylinder and instead of bouncing off were actually penetrating and knocking protons out. He had broken into and altered the atomic nucleus. Popularly it was said at the time that he had 'split the atom', but the effect proved to be less profound than that.

For a short time, indeed, it seemed that this might be the

beginning of a whole new experimental cycle in which nuclei could be broken apart by alpha particle bombardment and their composition established with confidence. The grail might be in reach. Rutherford was soon caught up in transferring to Cambridge and establishing himself at the Cavendish but when he resumed the experiments, often now in collaboration with Chadwick, he discovered that the dawn had been a false one. It seemed that only a handful of light elements could be affected by artificial disintegration of this kind and no great variety of outcomes was detected. They could force alpha particles into a few nuclei and knock the odd proton out of them, but no more than that. This was not splitting atoms but chipping them. Of course there were things that could be learned, notably further information about the size of nuclei and the strength of the protective shields surrounding them, but the Cavendish researchers were conscious that far from dismantling the nuclear machine they were merely taking out the odd screw.

Just at the moment when this disappointment was becoming clear, the two young men in Vienna started to report some rather different results. The first Rutherford heard of it was a letter from Pettersson in 1924 asking him to forward to the British journal *Nature* a communication claiming that he and Kirsch had pushed artificial disintegration to a new level. Not only had they disintegrated elements which the Cambridge men could not, but they believed they had spotted alpha particles as well as protons in the debris.

Rutherford's reaction was disbelief – he simply did not accept the results – and his first step was to try to kill the paper. To this end he wrote urging Pettersson to re-check his data and sent a second letter to Stefan Meyer, the director of the Vienna laboratory and an old friend. 'You know me well enough to appreciate that I would not interfere unless I thought the situation was serious,' he explained to Meyer.

I do not know Pettersson personally or his co-workers, but you do. He seems to me a man of originality and ingenious in his arrangements but I should judge he jumps to conclusions on insecure evidence. The subject of artificial disintegration is full of difficulties and pitfalls and wants investigators who are very careful in experiment and with good judgement. I am sorry to bother you over the matter but you will quite appreciate how important it is to Dr Pettersson and to your laboratory and also to the subject not to go along the wrong lines.[1]

The stratagem did not work, the Vienna findings were soon in print and despite Rutherford's reluctance to engage, matters deteriorated swiftly. He and Chadwick had bombarded many elements with alpha particles but were able to detect signs of disintegration – broken-off protons, in other words – only in nitrogen, boron, fluorine, sodium, aluminium and phosphorus. The Vienna team, however, now asserted that they were able to do much better, first breaking up the nuclei of silicon and then beryllium, magnesium and lithium – in fact as time passed it began to seem they could get results with almost everything they tried. Forced to respond, the Cavendish men returned to the laboratory and, using a slightly different experimental approach, discovered that they could indeed disintegrate two more elements, although not the same ones as their Vienna rivals. The response from Pettersson and Kirsch was a patronizing pat on the back and an announcement that they had now discovered disintegration effects in carbon. Perhaps, they suggested, the eminent Cambridge men should try using a better microscope to count their scintillations?

Despite further attempts at diplomacy by Rutherford things were coming to a breaking point.

It was Chadwick who took matters in hand, recognizing the fundamental nature of the challenge and returning to first principles. Since the two laboratories were using essentially the same experimental technique he saw that the source of these conflicting results must lie in the equipment, the working practices or the scientists themselves, so he subjected the routines and apparatus used at the Cavendish to a new scrutiny. Although these would still have been recognizable to Marsden from his 1909 experience and employed the same trusty combination of radioactive source, target and scintillation screen, by the mid-twenties things had become rather more elaborate.

Radium continued to be the standard source. Discovered by Pierre and Marie Curie in 1898, it was extremely rare and correspondingly expensive (when the women of America presented Marie Curie with a gram of radium for use in her experiments, the gesture cost them $100,000), but it had special qualities. In particular the alpha particles it emitted were tremendously effective projectiles, emerging as they did at speeds measured in thousands of kilometres per second. For years these projectiles made radium indispensable in an atomic laboratory and they yielded many important results, but inevitably over time demands increased. As the scientists strained to extend their knowledge the experiments became more complex and precise; more exotic effects were studied and finer beams were employed. The quality of the radium alpha particles remained more than sufficient – they were still fast and penetrating enough – but problems arose about the quantity. In these refined experiments the natural stream was sometimes barely enough to produce measurable results and the work became very slow and tedious. Patrick Blackett would note: 'I well remember spending hours in a darkened room with

Professor Rutherford, excitedly but rather wearily counting a very few weak scintillations in a microscope.'² It was like trying to conduct an opinion poll in the middle of a lonely Yorkshire moor: a very long time was needed to acquire even the barest minimum of data.

To compensate for the weakness in their particle source the scientists tried to increase the power and sophistication of their method of observation – they could get by with a sparing supply of projectiles, in other words, if they could be certain of logging every one over a sustained period. Thus, where Marsden had conducted his observations alone, in the mid-twenties a full-blown bombardment experiment in the Cavendish was likely to require three people. Two would go into a booth that was in as near to total darkness as possible and wait there while their eyes adjusted. The third, meanwhile, would put the final touches to the apparatus, close the shutters on the windows and pull down the blinds. When all was ready, the two would emerge from their box and take their places at twin microscopes. At a signal, one would start counting scintillations, sometimes by clicking a key like a Morse key and in other cases by making a mark with a pencil on a moving scroll of paper. This would go on for a minute, or perhaps until an agreed total had been reached, and then the other counter would take over and follow the same procedure, the two of them alternating for a maximum of one hour or until their eyes became tired, whichever came first. If anything went wrong with the apparatus in the meantime, or if some component needed to be changed as a control check, the two counters would be ushered back into their little dark booth while the third person turned the light on and made the adjustments.

Inevitably some people were more reliable counters than others. Chadwick was said to have an almost superhuman accuracy rate while Rutherford was a mediocre counter despite his long experience; he lacked the necessary patience and steady eye, and perhaps also the ability to empty his mind. Good counters, in fact, were so much in demand that every new crop of researchers at the Cavendish was tested and graded, and the better ones subsequently found themselves required every now and then to drop their own work and lend their eyes for some important series of counts. (This happened to one young student even after he had transferred to chemistry.) It could be dreadfully dull but equally it could be exciting to take part in a big experiment, especially as it carried the privilege of spending time chatting with senior scientists as they waited in the dark. Rutherford was always happy to talk about his ideas or to tap into his archive of anecdotes about great scientists, and for some students those spells listening to him before a count remained their happiest memories of the Cavendish.

Challenged by the Vienna researchers, Chadwick reviewed this whole system. As urged, he acquired a more sophisticated microscope and checked the results achieved with it against previous results, while at the same time he mounted a detailed study of the performance as counters of about thirty students, to establish the general level of reliability. At the end of 1926 he published his findings. 'We have good reason,' he concluded, 'for holding to the results obtained in this laboratory.'[3] A better microscope made no difference, while the counters gave reliable results up to certain clearly defined limits and all the key Rutherford–Chadwick disintegration results fell within those limits. It must be the Vienna researchers, he strongly implied, who were producing unsafe results.

The response was defiant: no fewer than three papers from Pettersson and Kirsch rebutting Chadwick and leaving no doubt they were satisfied it was the Cambridge results that were flawed. Renewed diplomacy brought Pettersson to the Cavendish in early 1927 but there was no resolution. That came only when Chadwick returned the visit, arriving in Austria at the end of the year. The atmosphere was tense and even bad-tempered but Chadwick was not to be deflected from a mission to test the Vienna method as rigorously as he had tested the Cambridge one. He was immediately surprised to find that the physicists did not count scintillations themselves nor did they rely on junior scientists; instead the job was done by three women specially employed for the task. 'Pettersson says the men get too bored with routine work and finally cannot see anything, while women can go on forever,' Chadwick reported to Rutherford.[4] There was nothing wrong with this in principle but when he watched these specialist observers in action and put them through a series of exercises he thought he detected a problem. They were not producing the clear-cut findings claimed for them. Inquiring further, he discovered that the women tended to know the character of the experiment in progress and were told what sort of results were expected. So he conducted a range of further tests in which the details of the experiment and its likely outcome were withheld. 'I ran them up and down the scale like a cat on a piano – but no more drastically than I would in our own experiments if I suspected any bias,' he wrote.[5] And with this the results suddenly changed. Not only did the data differ from the previous Vienna outcomes but they also now corresponded closely to the Cambridge ones.

In an unpleasant scene Chadwick confronted Pettersson and eventually convinced him that, although the mistake was an innocent one, the Vienna 'discoveries' were the result of suggestion rather than physical fact. Without meaning to, the observers were providing the scientists not with the correct

results but with the results they were hoping for. The tests left
no doubt. After all that had been said and published it was a
shocking professional humiliation for the Vienna team. Meyer,
the laboratory director, offered to issue a formal retraction in
a leading journal but remarkably Chadwick, acting on what
he believed would be Rutherford's wishes, declined the offer.
'I said no,' he recalled later, 'the best thing to do was to say
nothing more about it. They could drop the experiments and
say nothing.'[6] And that is what happened. For a short period
there was confusion in the scientific world as outsiders took
the controversy still to be 'live', but in the absence of any
follow-up work by Pettersson and Kirsch, or of any confirmation from elsewhere, their claims were eventually forgotten. The Vienna research was allowed to wither on the vine.

This episode, which had dragged on for four years, is
revealing in two ways. The first is the light it sheds on Rutherford. For all the volatility of his character, his experimental
rigour could not be faulted. He would not publish experimental results unless he was certain they were correct and
once he had achieved that certainty he could not easily be
shifted. And yet this confidence did not translate itself into
arrogance, for he maintained throughout the dispute the
anguished tone he had adopted in that first letter to Meyer. In
the same way he was relieved to have the whole affair buried.
'He would never have done anything which would have
caused pain to Stefan Meyer,' said Chadwick, 'but at the same
time he would never have indulged in this kind of public
argument to settle a matter of that kind.'[7]

Much more important, however, is the light shed on the
science itself. If the Vienna controversy proved anything it
was that the experimental technique that had sustained atomic
research for twenty years was approaching the limit of its
usefulness. On the one hand the radium emissions were barely

strong enough to yield results of any kind in the highly demanding work now under way and on the other the means of observation was no longer up to the task of providing data that could be trusted by everyone. Cambridge proved that its published results were accurate but in doing so it effectively demonstrated that the work could be taken very little further. Optical scintillation counting was subjective and would never be otherwise, and although various control systems could be employed they too depended on the human eye and brain. If the nucleus was to be probed with these techniques they would have to be pushed even further, but if that happened any findings that were produced would be more and more open to question. Rutherford and his colleagues, whether they were searching for the neutron, measuring the diameter of a nucleus or trying to provoke some new form of disintegration, needed tools capable of detecting and measuring effects that were ever fainter and more transient. Instead they were literally struggling in the dark. In ominous fashion the word 'unfortunately' was beginning to appear in their research papers and reports, as in 'unfortunately, the amount of disintegration is . . . very small', and 'unfortunately, on account of the small number of particles . . .'[8]

In 1927, therefore, the battleship of physics, if not exactly adrift, was losing momentum. The mysteries of the nucleus were as perplexing and beguiling as ever but the effort to crack it open and reveal its working parts was almost stalled. Rutherford had begun to flounder, to the point where a talk he gave on nuclear structure at a meeting in his own laboratory ended in chaos. 'The crowd fairly howled,' wrote a young student who was present. 'I think Rutherford came nearer to losing his nerve than he ever did before.'[9] Having committed twenty years of his career and the bulk of the resources of the world's leading physics laboratory to this one goal – sometimes in the teeth of carping from his peers – he was for the first time having to contemplate the possibility that he might not live to attain it. He was not, however, a man to give up. With characteristic fortitude he made a bold decision: if the old and trusted scientific method was failing him then he must have a new and better one.

4. A Way Forward

Among Rutherford's public honours and responsibilities in those years was the presidency of the Royal Society, the oldest scientific academy in continuous existence in the world. There is no loftier position in British science and Rutherford's predecessors included such men as Christopher Wren, Isaac Newton, Joseph Banks and Humphry Davy, and more recently William Crookes and J. J. Thomson. The members, known as Fellows, were the cream of their profession and in the 1920s the published *Proceedings* ranked among the world's most influential scientific journals. Each year at the end of November the Royal Society held (and still holds) what was known as the Anniversary Meeting, an AGM where necessary business was done, the society's prestigious medals were presented and the president made a speech. Tradition dictated that, besides bringing the Fellows up to date with what had been done on their behalf, the president could use the opportunity to air some views on a scientific subject of current interest, which in the way of things usually meant of interest *to him*. On the evening of 30 November 1927 Sir Ernest Rutherford did just that.

Rising to his feet in the main hall of Burlington House, the society's home, he looked out over the gathered Fellows – all in evening dress, as required, and all, in those days, men – and cleared his throat. He began with the housekeeping, first placing on record the deaths of various Fellows in the year gone by, among them zoologists, physiologists, mathematicians, chemists and one person who was not a scientist at all but a benefactor of science, the Earl of Iveagh. To each Rutherford paid tribute. He next reported on the financial state of the society, which appears to have been healthy although the burden of the publications was increasingly heavy. And he chivvied the Fellows about attending ordinary meetings: poor turnouts, occasional though they might be, were embarrassing to the society and discouraging to speakers, some of whom travelled long distances to present their papers. After this he paused, signalling a change of course. 'In the short time at my disposal,' he resumed, 'I would like to make a few remarks on the results of investigations carried out in recent years to produce intense magnetic fields and high voltages for general scientific purposes.'[1]

It was a research field in its infancy, he explained, but one of considerable promise and 'scientific men thus naturally follow with great interest advances in these directions'. He reviewed the state of the art. So far as high voltages were concerned, the need for X-ray machines for medical purposes had prompted the development of reliable electrical transformers for hospitals operating at up to 500,000 volts. This, however, was by no means the limit of achievement, for voltages as high as 5 million had been attained at the Carnegie Institution in Washington, DC, by the use of an apparatus known as a Tesla transformer. The effects were striking, with spectacular sparks, but the Tesla apparatus produced only transient peaks rather than sustained voltages, so its potential for use in the laboratory was limited. What Rutherford wanted to highlight was a third avenue of development which was being investigated in the power industry, particularly in the United States. 'In order to transmit electrical power economically over long distances, there is a tendency to raise the voltage in the transmission lines. This increase of the operating voltage has led to the need of very high voltages to test the insulating properties of these lines and their transformers and the effect of electric surges in them.' In other words, power companies had found they needed to generate very high voltages in their laboratories in order to test the equipment they wanted to use commercially for transmitting electricity from place to place. To do this they employed transformers in series, known as cascades, and had by these means achieved 2 million volts and more, with the possibility of 6 million on the horizon. Rutherford had seen such cascades at work and he reported that it was 'a striking sight, giving a torrent of sparks several yards in length and resembling a rapid succession of lightning flashes on a small scale'.

All of this was no doubt of some general interest to the membership of the Royal Society, and especially to the physicists, but it was essentially technology. It might make their domestic electricity supply more efficient in the long run but it would not in itself advance the frontiers of knowledge. Rutherford, however, had in mind a specific use to which these high voltages might be put.

From the purely scientific point of view interest is mainly centred on the application of these high potentials to vacuum tubes in order to obtain a copious supply of high-speed electrons and high-speed atoms. So far we have not yet succeeded in approaching, much less surpassing, the success of the radioactive elements in providing us with high-speed alpha particles and swift electrons. The alpha-particle from radium C is liberated with an energy of 7.6 million electron volts . . . the swiftest beta-rays [electrons] from radium have an energy of about three million electron volts, while a voltage of more than two million would be required to produce X-rays of the penetrating power of gamma rays.

Here was Rutherford's point: if the particles that came out naturally from radium were no longer adequate for his purposes in the laboratory, then maybe the time had come to look at ways of producing streams of fast particles artificially. And since the electricity industry was beginning to generate voltages of roughly the same order as the energy of natural fast particles, then perhaps those voltages should be employed in the task.

It would be of great scientific interest if it were possible in laboratory experiments to have a supply of electrons and atoms of matter in general, of which the individual energy of motion is greater even than that of the alpha particle. This would open up an extraordinarily interesting field of investigation which could not fail to give us information of great value, not only in the constitution and stability of atomic nuclei but in many other directions.

He acknowledged that difficulties lay in the way of such a development but said that it had long been his ambition to have such a tool at his disposal in the laboratory, and he was 'hopeful that I may yet have my wish fulfilled'.

While this connection that Rutherford made between electrical voltage and fast particles would have been readily understood by most of his listeners it is likely that a few of the Fellows, perhaps some of the zoologists and physiologists, were in the dark. Electricity is a manifestation of one of the fundamental forces of nature, of which two were known in the 1920s. Gravity, which pulls apples to the ground and holds our planet in orbit around the sun, had been identified in the seventeenth century by Newton, while it was Clerk Maxwell, the first director of the Cavendish, who in the nineteenth century defined and described the second basic force: electromagnetism. Like gravity this second force was constantly at work around us but it was much more powerful and, as we have seen, it was different in another way, for it came in distinct varieties. In magnetism these were the north and south poles and in electricity they were the positive and negative charges. In both cases a simple but all-important rule applied: opposites attracted each other while like charges or like poles repelled each other. Push the north pole of a magnet towards the south pole of another, for example, and they will jump together; push a north pole towards another north pole and they will jump apart. The same rule – likes repel and unlikes attract – applied to electrical charges. The alpha particle carried a positive charge and so did the nucleus, and when an alpha particle flew out of a radium nucleus during radioactive decay the force propelling it was the repulsion between the two like charges. No one had any idea how, when it was inside the nucleus, the alpha particle stayed there in the midst of all the other positive electricity, but what they did know was that whenever an alpha particle happened to cross the barrier, whenever it found itself immediately on the outside of the nucleus, the rule that like charges repel each other suddenly came into operation, and with extremely violent effect. As if hurled by a catapult, the positive alpha particle leapt away from the positive nucleus at tremendous speed. The scientific phenomenon that had supplied Rutherford with his fast projectiles for so long, therefore, was essentially electrical and obeyed familiar rules. It was also a naturally occurring phenomenon. But now he was fed up with hanging around in the dark waiting for nature to fire its own ammunition in its own time and he was saying that perhaps the technology of the electricity industry was sufficiently advanced to supply him with an alternative.

In essence what he wanted was a machine that would create an extremely powerful seat or focus of positive electricity in imitation of the radium nucleus, coupled with a second machine producing millions of positively charged alpha particles or protons. These projectiles would be released as close as possible to the high voltage, where they would experience the full repulsive force and immediately reel away at high speed. The new apparatus could then harness these fast particles to bombard whatever targets the scientist chose. It would be, in short, an artificial particle accelerator and it would be better than a radium source because unlike radium it would be adjustable. You could push the voltage up as high as the technology would allow, and over time that ceiling could be expected to rise considerably. You could also feed in as many particles as your apparatus would produce. You could set both the voltage and the flow of particles with precision to any level you wanted, to suit the particular needs of your experiment. Potentially such apparatus might even allow physicists to break up all atomic nuclei at will.

If the pure science of this proposition held few mysteries for physicists, the same could not be said of its practical application. The idea of taking equipment from the electrical industry that was capable of producing 5 or 10 million volts and installing it in a laboratory was about as far from the usual, do-it-yourself workbench practices of the time as could be imagined. It would cost sums of money of a kind that only the biggest and most profitable power and engineering companies could afford, sums far beyond the budgets of the richest university laboratories in Britain. Besides, a very large technical leap was implied. To create suitable conditions for acceleration the voltage would have to be applied in a controlled way, not as a sudden discharge or spark, but as electrical potential, held in a steady state. And what material would you use to contain such huge power? Glass, which resists electric current relatively well, had proved useful in laboratory work to date but in 1927 it was hard to believe that a glass vessel could be constructed that would be capable of withstanding millions of volts. Already, in fact, there were signs that the practical limit for glass was half a million, beyond which came sparking, shattering or a disruptive leakage of charge at the joints and edges. And voltages of this order were almost as unpredictable as lightning, so applying them to a glass tube and holding them there would not only be tremendously difficult, it would also be dangerous – a little like asking a lion tamer to teach tricks to half a dozen Tyrannosaurus Rexes at once.

In the language of a later age, Rutherford was indulging in 'blue sky' thinking, although it is possible that some in his audience thought it was something worse – desperate or even hare-brained. He was not at all embarrassed. He knew his speech would be published in the *Proceedings* of the society and read by scientists the world over and this was the message he wanted to send. Given the deadlock in efforts to comprehend the nucleus he felt this was the most promising way forward, and if he was going to have results in his lifetime then the sooner his fellow physicists began thinking about it the better. That personal note was unambiguous – 'I am hopeful that I may yet have my wish fulfilled.' And such a proposal was not quite without precedent; in fact Rutherford himself had mentioned the artificial acceleration of particles in print as early as 1913. Since he and Chadwick had moved to Cambridge in 1920, moreover, the younger man had suggested more than once that they should investigate the possibilities and had even attempted some modest experiments. 'But there wasn't the equipment available,' Chadwick remembered later, 'and to be quite frank I wasn't the man to do it.'[2] He was keen enough on the idea to press Rutherford but he got nowhere. Perhaps in the professor's mind the time was not ripe even for small-scale experiments, perhaps he was daunted by the cost or perhaps he had not yet acknowledged to himself the obsolescence of his existing techniques. By November 1927, however, he had clearly changed his view and in a rather grand way, for he was not speaking of putting a toe in the water of artificial acceleration, but envisaging the final product.

As for the economic and cultural leap implied in such work, Rutherford had already taken some steps in this direction, although not in the field of nuclear research. One of the most surprising aspects of the Cavendish story in the 1920s is the arrival and rapid rise of a young Russian, Peter Kapitza, whose brilliance and charm captivated Rutherford. He came as part of a visiting Soviet scientific mission in 1921 and simply stayed. An early success with experiments designed to reveal how alpha particles progressively lost their energy was followed by a switch into research with magnetism which grew steadily more ambitious. Although this work had little relevance to the nucleus and although it soon became very demanding of equipment and money Rutherford consistently backed it, and when, in that 1927 address to the Royal Society, he moved on from his discussion of high voltages to magnetism, it was mainly Kapitza's work he discussed. By the standards of the Cavendish a fortune had been invested in the ever more powerful machinery the Russian used to create his magnetic fields, most of the money coming from a special government grant negotiated by Rutherford. The outcome of the research does not concern us here, though it would eventually yield a Nobel prize, but the Kapitza story demonstrates that Rutherford was capable of embracing work that involved large-scale electrical engineering and comparably large financial investment. As things turned out, neither he nor physics itself was ready in 1927 for the 10-million-volt particle accelerator on a laboratory scale, but the climate was right to make a beginning.

A few days after the Royal Society speech, as December began and the end of his first term in Cambridge came into sight, Ernest Walton knocked on the professor's door. After six weeks or so the Irishman had completed his training in the Nursery and had come to see Rutherford to discuss his research project. He had not arrived at the Cavendish with a research idea in his head, indeed it is likely that when he turned up in October his mind was filled with hydrodynamics rather than atomic physics, as he still had that M.Sc. thesis to complete. But at some stage while he was working in the Nursery a thought came to him; it was only a thought but he liked it and when he was called into that scruffy office to say his piece he suggested it. This itself was unusual, so far as Rutherford was concerned. The usual practice was for the laboratory director to come up with the ideas, with help from Chadwick, not because they wanted to dictate what was done (although they certainly wanted to determine the general thrust of the research) but because few young students had good ideas suitable for Ph.D. study. And Walton's suggestion itself was novel, for he declared that he wanted to accelerate particles by electrical means. He always insisted afterwards, and there is no reason to doubt him, that he had no idea this was a subject Rutherford had just been discussing before the Royal Society; the notion had simply come to him. The timing, however, can only have helped and Rutherford instantly liked the idea.

What Walton proposed was not that he would undertake the enormous task of erecting 10-million-volt equipment in the Cavendish – such a project was unthinkable for a junior researcher – but to seek a short-cut to artificial acceleration using apparatus that would fit on a workbench. His plan was to whirl electrons around in a circular vessel, employing an arrangement of electrical and magnetic fields which would increase their speed with every revolution. If the electrons could be maintained in the circle long enough, then in principle they might reach energies higher even than natural alpha-ray levels, although only for very short periods. There was no evidence that electrons, slight as they were, would be capable of penetrating nuclei even at very high speeds – only alpha particles had done that – but it was worth trying, just to see what would happen, and in any case fast electrons would have other applications in atomic work. It was, moreover, just the sort of project to edge the laboratory in the direction indicated in the Royal Society speech. Rutherford instantly saw that Walton's own suggestion for how he might create the necessary fields was impractical, but this was no object since an alternative method sprang readily to mind. J. J. Thomson, though no longer director of the Cavendish, still conducted research there in a large room known as the Garage, and in some of his recent work with electrons he had employed apparatus that Walton could copy. That would give him a start.

Next Rutherford had to find Walton somewhere to put his ideas into action, and he knew just the place. Rising from his seat he led the way down the gloomy corridors and staircases, out across the inner courtyard, beyond one of the big undergraduate laboratories to a corner of the Cavendish that abutted the Department of Physical Chemistry. Going down a flight of stairs he threw open a door to reveal a large, spartan research room with plain brick walls, a bare wooden floor, an abundance of stray wires and some basic wooden fittings. It was a sort of basement, largely below ground level but with daylight coming in through short windows in the upper part of one wall, where the shadows of legs could be seen passing by in the street outside. The fittings included three workbenches, two of them occupied with apparatus and obviously in use, and the other one empty. Until recently this had belonged to an Australian student, Leslie Martin, but he had returned home after completing his research. Here Rutherford installed Walton, and in the couple of weeks remaining before the Christmas holiday the young Irishman laid the groundwork for his project and made the acquaintance of his neighbours.

He was surprised to discover that one of them was already working on accelerating electrons, although by a very different process. This was Thomas Allibone, a businesslike Yorkshireman of the same age as Walton who was known to friends as 'Bones'. Allibone had a background in industry, having worked for a spell in the research laboratories of Metropolitan-Vickers in Manchester, and it was he who had introduced the first real high-voltage studies to the Cavendish a year earlier. The story of his arrival is instructive. At Metro-Vick, as the great engineering combine was known, he had worked on the development of transformers and switching gear for the electricity industry and that gave him the idea of approaching Rutherford for the opportunity to apply this experience to atomic investigations. In particular he wanted to use a Tesla apparatus to accelerate electrons. The firm agreed to help with equipment and he was able to raise funds through grants and scholarships, so all he needed was Rutherford's go-ahead. The professor had been doubtful, partly because he 'didn't want anything dangerous on the premises', but he had eventually brought Allibone to this same room and asked him what voltages he might achieve in such a space. 'I don't think you could get more than half a million here,' replied the Yorkshireman, looking dubiously around. 'Half a million's no good,' sniffed Rutherford. 'It took me eight million to disintegrate nitrogen.'[3] (By this he meant the energy of natural alpha particles from radium.) None the less he encouraged Allibone, saying that even if there were no disintegrations the work would be useful for studying the behaviour of fast electrons. By the time Walton arrived, therefore, Allibone had been building his machine for a year, but progress was slow. Rutherford's interest, however, had grown considerably and there can be no doubt that discussions with Allibone about the potential for high-voltage work helped prompt his comments at the Royal Society.

The other scientist working in the room was John Douglas Cockcroft, although in practice he was not often to be found there. Cockcroft, who would eventually become Walton's research partner and the other central character in this story, was in several respects the most senior of the three. He came from the northern town of Todmorden on the Yorkshire–Lancashire border, was the oldest of five sons of a small-time mill-owner and had come to the Cavendish by a tortuous route. In 1927 he was thirty, and that six-year difference in age with Walton and Allibone carried a heavy significance, for while they had been schoolboys ten years earlier Cockcroft was a soldier on the Western Front. When the First World War broke out he had been a teenage student of mathematics at Manchester University (where he heard and was inspired by Rutherford's lectures) but in 1915 he dropped that, first to work with the YMCA supporting the war effort and then, when he was old enough, to enlist with the Royal Field Artillery. Without doubt he could have been an officer but he chose instead to volunteer as a rank-and-file gunner and was trained in the signals section, where he achieved 100 per cent in every discipline. He reached France just too late for the Battle of the Somme but spent well over a year in frontline service, often under fire and contending with lice and cold, shells and shrapnel and 'the devilish mechanical chatter' – as he put it – of the machine guns.⁴ He fought at Passchendaele and Cambrai and on one occasion was mentioned in dispatches after emerging as sole survivor from a forward observation post, having kept lines of communication open alone for thirty-six hours.

Throughout it all he wrote frequent, thoughtful letters home to his parents, his brothers and his school sweetheart, Elizabeth Crabtree, the daughter of another mill-owning family. He was modest about his involvement in the fighting and grateful he was never required, as the infantrymen were, to launch himself across no man's land in the teeth of enemy fire. He wrote of battle: 'Hell is not the word for its utter terror and awfulness. We are lucky in seeing little of it; what we do see is quite enough.'⁵ He did not dwell on this, though, and the letters usually had a reassuring tone. A fine day, a visit to the cathedral at Amiens, a rare hot bath or a chance encounter with another son of Todmorden – these and the letters and parcels from England kept his spirits up. And besides cakes and gloves those parcels often brought him books, because whenever army routine permitted he would read, not maths texts but fiction and the classics, from Shakespeare to Turgenev and from George Eliot to Machiavelli. He must have cut a curious figure for not only was he teetotal, from a relatively well-off background and interested in architecture and literature unlike most of the signallers, but he was also better equipped to grasp the technical side of gunnery than most of the officers. The anomaly was too great to last and it must have been a great relief both to him and his family when, during a spell of home leave in late 1917, he finally applied for a commission. Most of 1918 he spent far from the front, training in Northamptonshire and on Salisbury Plain, and when he finally passed out as a lieutenant it was just in time for the Armistice and demobilization.

From the age of eighteen to twenty-one, then, Cockcroft was at war, and when he picked up his education again it was not to study mathematics at university but electrical engineering at Manchester College of Technology, which he believed would prepare him better for paid work. From there he graduated to Metropolitan-Vickers, where the same enlightened management that was later to support Allibone recognized his potential and encouraged him to resume his academic studies. With the company's help he went to Cambridge, but not to the Cavendish; he studied maths. For such a technically minded student, however, the draw of physics, and in particular the magnetic power of Rutherford, was too strong to resist. With the professor's blessing he spent much of his spare time in the laboratory and as soon as he had his maths degree he enrolled at the Cavendish to take a Ph.D. That was 1924 and he was twenty-seven; very soon he was a fixture of the place.

It was not just that he brought with him expertise in electrical engineering and unusual contacts in industry; it was to do with his personality. Cockcroft was an exceptionally good listener, patiently hearing everything that was said and jotting things down in tiny spiderish script in a black notebook. Though he spoke sparingly (much later, his children would have a rule at home that Daddy could not leave the dinner table until he had uttered two whole sentences), what he said was invariably helpful and where his ideas involved taking on responsibilities he accepted them cheerfully and without fuss. He needed no supervision and he made life easier for others, and in a busy laboratory peopled by scientists who were often not very practical, such a man was invaluable. Rutherford trusted him, but more importantly he became in the professor's estimation an essential complement to the mercurial Peter

Kapitza, keeping the Russian's feet on the ground and ensuring that his expensive apparatus was built and used along sensible and economical lines. A good deal of the equipment for the magnetic work was built for Kapitza by Metro-Vick in Manchester and it was Cockcroft who maintained this liaison, kept the accounts and oversaw the installation – having grown up with big machines at the mill he had no fear of them. So deep was his involvement that within a year of his arrival Rutherford was applying to the government for a special annual grant of £100 to pay him for this work. 'Cockcroft is a very useful man,' he wrote, and Kapitza echoed the view.[6]

Like most useful people he became very busy indeed and by 1927 it seems that he could spare little time for his own research. As it happened this was no great loss, for as an experimenter he had shortcomings. His project was not a nuclear one but related to the interaction of molecules and metal surfaces, a topic with a less fundamental and more applied character. For this he did not need large apparatus like that used by Allibone but a benchtop set-up involving a fair amount of glass tubing under vacuum. Although he had learned some glassblowing in his spell in the Nursery he had no aptitude for it so the tubes were constantly cracking and Allibone, whose schoolteacher father had taught him to blow glass as a boy, was frequently called upon to help. Cockcroft was also, if not quite absent-minded, then preoccupied, as Allibone recalled:

John would come in early in the morning – well, not too early – switch on pumps etcetera for his metal vapour deposition apparatus and then dash out to Kapitza's laboratory or to St John's [his college] or elsewhere, completely forgetful of the need to turn on the water or something else, and one or other of us would find his apparatus in a critical state just before it disintegrated.[7]

These then, were the three occupants of the big room. Allibone took up the greatest space as he played around with his half-million volts. His equipment, including a two-metre-high Bakelite tube which he had personally escorted from the station in a taxi, steadily took shape and could soon produce sparks a metre long. At Rutherford's instigation he conducted a test, supervised by a professor from the department of medicine, in which a tame rat was electrocuted, so the dangers were clear. Cockcroft, though an erratic visitor, was always a welcome one for he proved just as useful to his neighbours as he was to everyone else, giving advice and tracking down bits and pieces of equipment that they needed. Walton soon settled down in this company and fitted in well. His own character struck a curious balance. He did not lack confidence in his own abilities or ideas but he was not someone to press them on others. This was not shyness but seems to have reflected a consciousness of his own standing relative to the scientists he found around him – they were in a different league from most of those he had known in Dublin. He was a young researcher with a job to do and his priority was to get on with it rather than make himself known. This attitude changed very little in the years to come.

Work on his own project did not begin in earnest until the new year of 1928 and he soon found himself in difficulty. The success of the scheme depended upon applying the electrical and magnetic fields to his circular chamber in such a way that the electrons would move in a stable orbit. Creating the correct conditions proved to be troublesome in both theory and practice and between January and May Walton designed three different sets of apparatus, first using a glass chamber and later a copper one. In none of them did he find the slightest evidence of fast-moving electrons. He suspected that his electrons were striking the walls of the chamber and losing their direction and that the culprits were probably stray electrical fields within the apparatus. Try as he might, he could not find a way of eliminating these and so in the middle of May he decided to abandon the effort.

All was not lost. He was preparing a theoretical paper which he felt would show that the basic principle was sound even if the technical conditions were not ripe in 1928. (In this he was correct, for just such a machine – it was called a 'betatron' – was developed a dozen years later.) And he already had an alternative experimental idea which he put to Rutherford and which, once again, the professor instantly liked. Instead of whirling his particles around he now intended to accelerate them in a straight line by passing them along a succession of cylinders. The gaps between these cylinders would be subject to electrical fields arranged in such a way that they would cause the particles to jump forwards. Each successive jump would increase the speed of the particle. Rutherford listened carefully, made a couple of calculations on the spot and then told him to go ahead. Working rapidly, the young Irishman designed and soon produced this linear apparatus but once again, when he tested it, he could detect no positive results. Returning to the drawing board he sketched a bigger and more elaborate version, but this required specialist craftsmanship and was a job for the technicians in the laboratory workshop. This meant he would have to wait. The months were passing; it was now late in 1928, and at that moment some news arrived in Cambridge which prompted a reassessment of the whole business of accelerating particles.

5. A Man in White Trousers

The news was unexpected in every way. It was a revolutionary idea built upon other revolutionary ideas and it arrived from the world of theoretical physics by a route that led from Leningrad through Copenhagen by way of Göttingen in Germany. Its source was a young man from the Ukraine, an unknown scientist called George Gamow, who would quickly become a great friend of the Cavendish.

Gamow was born in Odessa in 1904, into a family that was as colourful as he himself would prove to be. His paternal connections were mostly military – three uncles died in the Russo–Japanese war – but his father was a schoolteacher among whose pupils had been the young Lev Bronstein, later Trotsky. Close relatives on his mother's side included an Orthodox archbishop, a battleship commander and an astronomer who was hanged for plotting to assassinate the Russian prime minister. The mother died when Gamow was small so a bookish father was the dominant influence in his upbringing, although the son's later account suggests that he had little need of guidance. At six years old he observed Halley's comet from the roof of his home and at seven he was reading Jules Verne and experimenting with electric batteries. Soon he was mastering Euclidean geometry and in his early teens he smuggled Communion bread home to test the doctrine of transubstantiation under his microscope. ('I think this was the experiment which made me a scientist,' he would write later.)[1] He was evidently a prodigy. When the time came he went to the local university to study his beloved physics but soon found the course there no match for his abilities. Teachers and equipment were in short supply in Odessa after all the years of war and revolution, so George persuaded his father to sell the family silver and send him somewhere better: Petrograd.

He was determined to make a research career in atomic physics – 'Already in Odessa I was interested in the work of Rutherford.'[2] Paying his way by working part-time as an instructor in the Red Army (he taught physics to artillery cadets and had the rank and uniform of a colonel), he coasted through the undergraduate course, skipping most of his lectures, and secured his primary degree in 1925. Much of his time was spent in the company of a coterie of like-minded students who became known, for their bohemian ways, as the Jazz Band. They went to the cinema to watch the Hollywood silents, played tennis and parlour games – a lifelong passion for Gamow – and lounged in the easy chairs of the physics library discussing the latest work in their field. Against the background of revolutionary Leningrad (the city was renamed in 1924) they must have cut quite a dash. Though his interest lay in theory, Gamow chose an experimental project for his postgraduate research because, he said later, the experimenters were given rooms of their own and places to hang their coats, while the theorists were not. The choice may have suited his comfort but as we shall see it did not suit his talents. Fortunately, thanks to his avid reading of the journals and the discussions among his Jazz Band friends, he kept abreast of the remarkable developments then taking place in the field of theory.

Looking back from a distance of three-quarters of a century, it seems that in the mid-1920s two worlds of atomic physics existed almost independently. One was the world of Rutherford and the hard-headed experimentalists, to be found not only in Britain but also increasingly in the United States and in laboratories in Berlin, Paris, Leiden and elsewhere. The other was the world of the theorists, a predominantly continental and even Germanic domain. The theorists did not work with fixed apparatus but with ideas and argument, and their existence had a fluid, nomadic quality. Formally they would be scattered through the old universities in little groups, each clustered around a professor of reputation, but in practice they were often on the road, shifting freely from cluster to cluster. A bright young postgraduate would spend a term in Leipzig and a year in Zurich, with holiday courses in Vienna and Heidelberg bridging the gaps – visiting as many centres as he could to learn at the feet of the great men. The professors themselves also moved about, attending conferences, changing posts or merely paying visits to old colleagues. Of course their world overlapped with that of the experimentalists – they met sometimes and read each other's papers – but there was far less contact, less partnership, than we would expect. Their preoccupations were for a time very different; in particular, the theorists were struggling to meet a challenge that seemed to them absolutely fundamental.

The great achievement of their discipline, the foundations of which were laid in the days of Newton, had been to describe the laws that governed nature. With wonderful consistency, it had been found, the movement and interaction of objects could be laid out in mathematical terms which, when refined, yielded a small number of rules and formulae which appeared to account for virtually all the observable universe. Heat and light were reduced to algebra, and as we have seen in the nineteenth century new laws were written that satisfactorily described electricity and magnetism. So successful was this approach that, setting aside the matter of life itself, it was felt that the clockwork which ran our world was largely understood and people began to say that the only task remaining for physicists was the humdrum one of measuring its operation more and more exactly. Do not study that subject, young men were told, because it has all been worked out. Then, as the twentieth century began, serious difficulties arose with the 'classical laws'. In the 1920s theoretical physics was still struggling with these difficulties and one of the principal points of argument was what went on inside the atom.

Soon after Rutherford unveiled the nucleus in 1911 he was joined in Manchester by a young Dane, Niels Bohr, who would take up what was then the only university post in Britain explicitly devoted to theoretical physics. A gauche character with a huge head and a rustic air, Bohr was a profound thinker greatly admired – loved, even – by Rutherford, and like Rutherford he became fascinated by the nuclear atom. What intrigued him most was not the inner structure of the nucleus but the rest of the package, and in particular the behaviour of electrons. This was of great importance, not least because of its role in enabling atoms to connect with one another and form molecules. Bohr could see that from a theorist's perspective Rutherford's 'solar system' atom had a serious flaw, which was that under the classical laws of physics it was bound to collapse. If the electrons simply orbited the nucleus they must constantly burn up energy, and if they did that they would very soon become exhausted and spiral down to the nucleus. Since they obviously did not do this something else had to be going on, and to his considerable discomfort Bohr found that there was no possible arrangement which would conform to the classical laws.

This was not the first time that the laws had been found wanting. In the late nineteenth century scientists had come across a handful of anomalies where experimental results did not seem to match established theory, but their response had usually been to doubt the experiment or to blame prevailing ignorance; the laws were too successful to be called lightly into question. As the century drew to its close a German theorist called Max Planck took up one of these anomalies, relating to a phenomenon called 'black body radiation', and in 1900 he found a way of accounting for it mathematically. The equation balanced and the sums added up, but Planck's technique had a significant drawback: it worked only if you set aside an essential component of the classical laws, the principle of continuity. Continuity has an analogy in the kitchen: in baking, milk is 'continuous' in the sense that any given amount may be measured out and added to the mixture, while eggs tend to be 'discontinuous' – it is a perverse cook-book that asks you to separate one-quarter of an egg. In physics the classical laws considered light, motion, heat and energy generally to be like milk rather than eggs; they could come in any quantity, no matter how small. When atoms absorbed energy or gave it out, therefore, they must obey this principle and deal in infinitely variable quantities. Not so, said Planck. Black body radiation could be explained satisfactorily only if it was assumed that atoms dealt with energy in discrete and regular amounts. It was not a case of milk but of eggs – or in Planck's terms 'quanta', for this was the birth of the quantum theory. And Planck provided the essential mathematical tool or 'constant' to be used when calculating the size of the quantum, giving it the shorthand designation h.

Here was the first inkling of a sensational possibility: that the laws of physics that applied in the observable world might not, after all, be valid at the atomic level. Planck himself was dismayed at this prospect and his contemporaries greeted his idea with sullen scepticism, but in the years that followed the supporting evidence slowly accumulated. The most powerful came in 1905 from Albert Einstein, both in direct and indirect ways. Indirectly, his special theory of relativity contained another suggestion that the classical laws were not universal, although that theory dealt not with very small things but with very fast ones: conditions close to the speed of light. More directly, and on the small scale, Einstein's separate work on the so-called photoelectric effect employed Planck's quanta to show another circumstance where light must behave like eggs and not milk. The photoelectric effect occurs when light falling on certain metals displaces electrons from their surface and Einstein's explanation for this was that the light was behaving as a particle, physically knocking the electrons out of place. One whole light quantum – one egg of light – carried enough clout to shift a single electron. No smaller quantity of light could do the job, he insisted, and in fact no smaller quantity of light could even be said to exist.

While Planck's idea had been greeted almost with silence, Einstein's provoked uproar. In 1905, after all, any competent scientist could set up an apparatus that would prove beyond doubt that light travelled as a wave. Waves were continuous, steadily dispersing themselves as they grew wider and wider. They were not discrete particles moving in straight lines and maintaining their integrity as they went. So not only were Einstein and Planck challenging the laws of classical physics but they also seemed to be casting doubt on an experiment so basic that it could feature in undergraduate practical examinations. Yet within a few years experimentalists were able to report results which supported the new theory – beyond all doubt, they found, there were indeed circumstances in which light came in quanta. There was no avoiding it: light was made of waves, but it was also made of particles.

This was not a case of a new, correct idea supplanting an old, incorrect one. Convenient as it might have been, it was not possible to set aside the wave theory of light for the good reason that it was incontrovertibly, demonstrably true. And yet it seemed equally impossible to reconcile it with the quantum theory, which was now also demonstrably true. One British professor summed up the mood of confusion when he declared he would teach his students classical wave theory on Mondays, Wednesdays and Fridays and the new quantum theory on Tuesdays, Thursdays and Saturdays.

In 1911 Bohr, troubled by the electron orbits in Rutherford's nuclear atom and emboldened by Planck and Einstein, began to think radical thoughts. There was another way of looking at the orbits, he saw, which would satisfy the energy equations without requiring the atom to collapse, but once again it meant setting aside the idea of continuity. Instead of circling the nucleus in a steady, free and even way the electrons must have a limited range of possible orbits and every now and then they must jump from one to another. Each leap would involve either a release or a capture of energy, with a release bringing the electron closer to the nucleus and a capture taking it further away, and when Bohr looked at the amount of energy required for such a leap he was not surprised to find it could be calculated only with the help of Planck's constant, h. The leaps, in other words, were measured in eggs – in fact they were quantum leaps. Over the next ten years or so this Rutherford–Bohr atomic model, incorporating the leaps, gained general acceptance, bringing with it a wider approval of the quantum idea and advancing the recognition that the classical laws must have limitations. But Bohr had not solved the whole problem. For one thing his mathematics was neither pure quantum nor pure classical but a mixture of the two, and he himself was quick to point out that this was far from satisfactory. For another, Bohr had produced a mathematical interpretation of only a single kind of atom and that was the simplest of them all: hydrogen. Although his formulae seemed to work for one single electron orbiting one single proton, the business of extending them even to the next lightest atom, helium, was extremely troublesome.

Such was the climate that prevailed in theoretical physics when Gamow and his Jazz Band friends began hanging around together at Leningrad University. The classical laws had been challenged in several ways but the jury was still out and in the meantime there seemed to be no satisfactory way for theorists to get to grips with the workings of the atom. By international standards Leningrad was not a cutting-edge university in this field, but neither was it a backwater like Trinity College, Dublin, where Walton was then completing his degree. It had a good school of physics with a strong tradition and though on the theoretical side there were no world leaders on the staff the mood was outward-looking, the foreign journals were read and the discussion of new ideas was lively. Gamow and his friends were familiar with Rutherford's artificial disintegration work, with Bohr's atomic model and with the conundrum of light quanta, so when something fresh came along in 1925 they were as ready for it as anyone else.

A young man called Werner Heisenberg, who was based at the university of Göttingen but had also studied at Bohr's new institute in Copenhagen, suggested rethinking the atom all over again. What he proposed was not a new model for the atomic structure but an approach which did away with models altogether. Far too much time was being spent trying to describe the movements of electrons, Heisenberg believed, when this was something that might be beyond the capacity of human language to describe or even of the human mind to conceptualize. Instead, theorists needed to concentrate on what they could do with the known facts about electrons. This is what Heisenberg did and he soon found a promising way forward involving the use of some arcane mathematical techniques. Plotting and arranging the known information, he constructed a series of interwoven matrices which could be used to establish, not so much the facts about atomic behaviour, and in particular about electron behaviour, as the probabilities. And at the heart of this matrix system, once again, was *h*.

It was a bold departure and, after some fine-tuning by Heisenberg's colleagues at Göttingen, it soon began to prove quite successful, giving a range of results conforming with experimental evidence. But before anyone had time to master it properly another series of papers began to appear which suggested a quite different way forward. These were the work of Erwin Schrödinger, an Austrian professor who had found his inspiration in a little-noticed article by a French aristocrat, Louis de Broglie. De Broglie had asked a simple but original question: if (as in the case of light) things we think of as waves can be particles, was it possible that things we think of as particles could be waves? In particular, was it possible that the electrons in an atom behaved as waves? Doing the calculations, and once again making generous use of *h*, de Broglie found that what had been thought of as electron orbits was perfectly consistent with wave movement – a wave of the appropriate kind would travel around a circular canal in just the same way.

Physicists reading this for the first time were entitled to sympathize with Lewis Carroll's White Queen, who thought six impossible things before breakfast. After all, the electron was an old friend, no less firmly established in their minds as a particle than light had been as waves; they even knew its weight. But Schrödinger took up de Broglie's idea and pushed it much, much further, for what he presented in his papers in 1926 was nothing less than a mathematical key to the atom founded entirely on the premise that electrons behaved as waves. With a series of relatively simple equations he said he could do the same things that Heisenberg claimed for his complex matrix system, but more conveniently. Even the cleverest theoreticians were amazed, not least because, as Gamow would write later, the two procedures 'looked as different as a chicken fence and a pond'.[3] But sure enough, when Schrödinger's technique was put to the test it worked just as well, and since the mathematics was more familiar it was generally preferred.

The emergence of these two systems was widely taken as a vindication of Heisenberg's assertion that it was a waste of time to try to deduce anything about parts of the electron's existence which could not be observed. Sometimes it might be a wave and sometimes it might be a particle; more likely it was never quite either, but occasionally resembled one. Language was simply not sufficient for the task, and as if to prove it one leading scientist began to write light-heartedly about 'wavicles'. Heisenberg pushed this notion of the ultimate unknowability of the sub-atomic world one step further when he pointed out that it was impossible to establish the exact position of an electron and its exact velocity at the same time. You could know one or the other, but not both. This was the famous uncertainty principle.

If it appears by now that the scientists were merely surrendering to their own ignorance the impression is a false one, for the new 'quantum mechanics', as the mathematical systems were known, offered a very practical means of describing and anticipating electron behaviour. The whole approach bears some resemblance to the way we think about lotteries and raffles. In a lottery no one asks why an individual person has won the jackpot prize; we accept that the random process has operated and that this person has simply been lucky. If we want a more sophisticated understanding of what has happened the best we can do is to look at the lottery in a statistical way. We can find out the total of tickets sold, the numbering system used and the other details of the process and from these we can calculate the odds against any given ticket holder winning a jackpot. This still won't tell us why one particular person has won but it will at least explain the probabilities. In a similar way quantum mechanics was useless at plotting the path of an individual electron but by drawing on the known statistical data it could give an indication of the probability of electrons doing any particular thing. This was quite unlike the comfortable certainties of the classical laws, but in a field where things invariably came in millions or billions probabilities could be a highly effective tool.

While the new ideas stirred up intellectual turmoil across Europe, in Russia George Gamow was reaching a dead end in the laboratory. Again and again the photographic plates which were supposed to provide his experimental results showed nothing at all. The explanation was simple, though he did not learn it until later: his camera equipment, which was foreign made, needed to operate at 'room temperature' and it had not occurred to him that the chilly sub-climate of a Leningrad university building might not be what the manufacturers had in mind. His plates, therefore, were always underexposed. He abandoned the work and was given instead a problem in the field of theory but he made no progress with that either, possibly because it was pre-quantum theory and he was no longer interested. Then early in 1928 his superiors had a better idea about what should be done with this clever but somewhat feckless young man: they would send him abroad.

It was one of the triumphs of the physics community in the new Soviet Union that they managed to keep open their lines of communication with the outside world, and the principal figure behind that triumph was Abram Joffe, director of the Physico-Technical Institute in Leningrad and a scientist of

international reputation. He and his senior colleagues convinced the new communist leadership first that advanced
science, even theoretical science, was important to the future
of the revolution and second that it required the interchange
of ideas, even if that meant maintaining contact with the
capitalist world. It was on this basis that Joffe led a delegation
touring European laboratories and factories in 1921, the trip
which first brought Peter Kapitza to the Cavendish. Visits and
exchanges were kept up throughout the decade, although
given the sparseness of Soviet resources it was always a great
privilege for a scientist to travel abroad. That Gamow was
selected for such a trip is a measure of the potential that was
seen in him. The suggestion had come from Orest Danilovich
Khvolson, a retired professor who had taught Gamow undergraduate physics and then kept in touch. Khvolson had the
idea that the young man would benefit from attending the
summer school at Göttingen and he still had sufficient influence to fix it. Applications were made, visas were acquired, a
passport was issued and – most difficult of all – the vital permit
was granted to exchange roubles for marks. Whether the idea
came from Khvolson or Gamow or from someone else we do
not know, but at this stage the decision was also taken that
while abroad he should pay a visit to Bohr's institute in
Copenhagen. For this purpose Joffe, who corresponded with
Bohr, supplied the young man with a letter of introduction
asking his old friend to let this '*russische theoretischen Physiker*'
attend a few of the institute's seminars.[4] And so in the first
days of June George Gamow, lanky, blond and wearing white
tennis trousers because they were the only decent pair he had,
stepped aboard a steamer in Leningrad, waved goodbye to
a flock of adoring friends on the dockside and sailed for
Germany.

The story of his arrival in Göttingen a few days later gives a hint of how lightly he carried the burden of revolutionary expectation that had been placed upon his shoulders. It was late afternoon when he reached the medieval city in deepest Saxony and checked his suitcase in at the station. One other Russian physicist, Vladimir Fok, had been given permission to attend the summer school and he had arrived a few days earlier. Since Fok was the only person Gamow knew in town he promptly tracked him down, locating him just as he was leaving for a party arranged by no less a person than Max Born, the director of the Institute of Theoretical Physics at Göttingen. 'Come along,' said Fok. 'I am sure Professor Born will be glad to see you.'[5] No doubt still sporting his white trousers, Gamow tagged along and found himself in the midst of a thoroughly jolly occasion – 'a typical German party, with dancing and such games as using soup spoons to pick up potatoes arranged on the floor or attempting to capture, with one's teeth alone, apples floating in a bucket'.[6] Nothing could have been more to his taste, particularly as 'there was also a reasonable amount of Münchner beer, and a divine liqueur called Kloster Geist'.[7] By the time the party ended he was in company with a young woman student and, taking a Chinese lantern to light the way, he walked her home. We can only guess at his intentions but we know the outcome. Reaching her doorstep the woman bade her escort a firm goodnight and left him there, lost in a strange town, without a suitcase or a bed for the night and perhaps a little the worse for Kloster Geist. He found emergency accommodation in what turned out to be a brothel, so the next morning his first task was to seek proper lodgings and these he found in the home of a professor's widow, where he would sleep in the more fitting surroundings of the old man's library.

The summer school was a great occasion. Heisenberg was not there but many others were, and in the lecture halls and seminar rooms as well as on street corners and in cafés the talk was all of the new theories. Gamow preferred the informal to the formal and soon found like-minded friends in Eugene Wigner, a Hungarian who worked in Göttingen, and Fritz Houtermans, an Austrian with a vast repertoire of Jewish jokes and stories. Brought up in Vienna's Jewish intellectual circles, Houtermans was very left-wing and by Gamow's account was drawn to Russians by curiosity about the progress of communism. Being Viennese he had quickly located the best coffee shop in town, a Konditorei five minutes from the main physics building, and there he ensconced himself in a corner and drank coffee hour upon hour, the empty cups filling the table in widening circles around him.

As usual Gamow skipped the lectures but he was as active as anyone in the scattered debates and they soon convinced him of something he had begun to suspect before leaving Russia. Too many people were busily trying to plot the activities of electrons and the mathematics was becoming very complicated. He did not like working in a crowd and, characteristically, he was not attracted by lengthy calculations, so he started to think about doing something of his own and the idea came to him of applying the new techniques to the nucleus. It might seem odd that nobody had done this before and that the initiative should have been taken by an unknown twenty-four-year-old Russian, but it should really be no surprise. There had been very little time and the problems relating to electrons had been mesmerizing; what was more, the nucleus remained so mysterious and elusive that most theorists thought it a problem for another day, a day when the experimenters had provided more in the way of data. Gamow, in short, was attempting something that most of his fellows thought pointless.

To make a start he needed to establish the existing state of knowledge, so he went to the Göttingen physics library and consulted back numbers of the *Philosophical Magazine*, the British journal in which Bohr had published his papers on atomic structure and to which Ernest Rutherford also contributed. In the volumes for the previous year, 1927, Gamow found a paper by Rutherford entitled 'The Structure of the Radioactive Atom and the Origin of the Alpha Rays'. An attempt to take stock of the research of the 1920s and make some guesses on the basis of what had been discovered, this article cost Rutherford great effort to write and went down very badly in Cambridge. It was the same work that had been howled down by the students, and both Chadwick and J. J. Thomson had made clear their low opinion of it. For Gamow the interesting part was a relatively minor one, Rutherford's description of the process by which he believed alpha particles escaped from the nucleus in the course of the radioactive process. This was a puzzling matter because it was well known that the nucleus was surrounded by an electrical barrier of great strength which both shielded it against attack from the outside and kept the contents packed inside. In principle it did not seem that alpha particles had the energy to cross this barrier. Rutherford's explanation was ingenious: it involved electrons travelling to the outer limit of the nucleus with the alpha particle and then, having helped it on its way, falling back inside – like tugs pulling a liner out to sea and then returning to harbour.

The more Gamow read of this theory the more certain he was that the whole approach was misguided and before he had even closed the journal he had his inspiration. This was, he realized, exactly the sort of problem that quantum mechanics was meant to solve. The *raison d'être* of the new technique, after all, had been to account for sudden changes in state of electrons – the quantum leaps. At one moment they were in this orbit and then the next they were in that orbit; what rules governed such changes? Now Gamow was looking at something quite similar: one moment the alpha particle was inside the nuclear energy barrier and the next it was outside. Rutherford's tug-boats, he thought, were laughably old-fashioned, but what did quantum mechanics suggest? First of all it did not need to visualize. According to the rules the change from one state to another simply occurred and any attempt to explain it from human experience was at best pointless and at worst obstructive. Second, this was a matter of probabilities and with luck those probabilities might be calculated. All Gamow needed were Schrödinger's wave equation, with which he was familiar, and sufficient data about the various distances and energy levels around the nuclear barrier. He threw himself into a search of Rutherford's various disintegration papers to see how much information was there and soon he had what he wanted. Then he retreated to his book-lined room in the widow's house, sat down and wrote a quantum mechanical theory of radioactive escape.

Let us break the rules and visualize. The nucleus is like a cereal bowl, with high sides representing the energy barrier. Lying at the bottom of the bowl are several marbles representing the components of the nucleus. Some are alpha particles. The marbles, we know, cannot leave the bowl because they do not possess on their own the energy to roll up and over the side. But whether we like it or not some of the marbles do occasionally turn up outside the bowl (it may help a little if we remember that these marbles may behave as waves when they choose). For convenience we may call this transition 'leaking' or 'tunnelling' but in reality it is neither; the marbles are inside at one moment and outside the next, and that is that. Quantum mechanics offers the means to quantify and anticipate such transformations and Gamow now used it to write the formula. It was not even very difficult to do by the standards of theoretical physics, although at one point it did become too much for Gamow. (A mathematician friend supplied the answer, observing tartly that he would fail any undergraduate stumped by the same problem.) The next step was to try it out on some other theoreticians so, discreetly, he consulted Wigner, the Hungarian, who declared himself impressed, and Houtermans, the Viennese, who liked the idea too. That was good enough for Gamow and he returned to his study to polish it up for publication. This took him until 29 July, when he signed his name to the typescript and put it in the post.

In terms of the history of physics this was a milestone. By applying quantum mechanics to the nucleus Gamow had opened the door to a new understanding of the kernel of matter and his idea would resonate through the years to come, most notably providing Rutherford and the Cavendish with a chance to escape from their nuclear impasse. Unfortunately for the young Russian, however, things moved in more mysterious ways than he could have expected. He had sent the article to the journal *Zeitschrift für Physik*, which was for him the natural choice. It was customary for Russian theorists to publish in the German journals and in fact Gamow had already co-written a minor paper for the *Zeitschrift*; more than that, he was in Germany and the dateline on his paper was Göttingen, so it was doubly appropriate. But the *Zeitschrift* was a monthly and as one of the leaders in its field no doubt it had quite a backlog of papers; there was bound to be a delay, especially with an article by an obscure Russian whose work arrived without endorsement from a prominent academic personality. In the circumstances it is impressive that Gamow's theory made it into print on 12 October.

In that interval of ten weeks a great deal happened. The

Göttingen summer school came to an end in July and it had been Gamow's intention to travel north to visit Bohr's institute in Copenhagen. From his fellow-students at Göttingen, however, he learned that the Bohr institute would be closed for most of August so there was no point in being there then. Faced with a choice between returning early to Leningrad and hanging around in Germany until September, when he could go to Copenhagen, Gamow (it should be no surprise) chose the latter. That he did not have enough foreign money for this does not seem to have concerned him. He therefore wrote to Bohr explaining the delay and enclosing his letter of introduction from Joffe, and a cheerful reply soon arrived to say that he was welcome to come at the later date. Gamow had a month to kill and he spent most of it writing, with Houtermans, a follow-up article for *Zeitschrift* giving more mathematical detail on his theory. It was Houtermans who insisted this was necessary and the paper was largely written among the coffee cups in the Konditorei, although they made a few trips to the mathematics building to use an electrical calculating machine. The Russian would later grumble that it was all very tedious.

By the time Gamow reached the Copenhagen institute on 27 August and asked to see Bohr he had barely enough money in his pockets to see him through another day, so when the secretary told him the professor would not be available for a few days he had no choice but to throw himself on her mercy. She took pity and an appointment was fixed for the same afternoon. In 1928 Niels Bohr was forty-two years old, a Nobel winner and a Danish national hero. In the world of physics he held a position of unique influence and his institute was emerging as the foremost centre of advanced theoretical thinking in the world. Though he had not personally put forward the main ideas in the new quantum mechanics Bohr had played a leading part in the great debates they provoked, including a celebrated public duel with Einstein just a year before, and he was generally regarded as the outstanding philosopher of the subject. It was only natural, therefore, that Gamow should be so keen to spend a few weeks in this man's company. Meeting Bohr at last he laid out his wares, giving a full account of his new theory of radioactive escape and explaining that his paper was awaiting publication at *Zeitschrift*. The professor was both impressed and intrigued, and he wasted no time. 'My secretary told me that you have only enough money to stay here for a day,' he said. 'If I arrange for you a Carlsberg Fellowship at the Royal Danish Academy of Sciences, would you stay for one year?'[8] They shook on it there and then.

6. A Finite Probability

The gods on Olympus did not have a better time than the theoreticians who studied in the Copenhagen institute in the 1920s and 1930s. Bohr's standing and friendship attracted the most illustrious visitors: Heisenberg was a particular favourite; Schrödinger stayed in 1927, at the height of his glory; the acerbic Wolfgang Pauli, author of the Pauli exclusion principle, returned again and again; from England came Paul Dirac, whose equations accounted for the spin of electrons, and from the Netherlands Paul Ehrenfest, a father figure to many in physics. Such men as these – and Bohr himself – provided stimulus and inspiration for the rest, a roll-call of the brightest young theoreticians of the time, drawn from all over Europe and beyond. The conditions were ideal: Bohr was indifferent to matters of discipline and punctuality so the students came and went as they pleased; seminars could continue late into the evenings until all were exhausted; pride and vanity were not tolerated and play was regarded as essential. George Gamow had truly fallen on his feet.

He was among the first in place for the autumn term of 1928 but in the next week or so the others drifted in. Among them were two young Cambridge men, Nevill Mott and Douglas Hartree. The Cavendish did not, strictly speaking, have a theoretical department but an informal one had developed thanks to the close friendship between Rutherford and Ralph Fowler, a mathematician married to Rutherford's daughter, Eileen. Fowler took an interest in atomic problems and encouraged his students to follow suit, and so had become leader to a small group of Cambridge theorists. Although there were no facilities for them at the Cavendish they would spend time there chatting with the experimenters, attending meetings and discussing problems in the little library. Mott and Hartree were members of this group and they had come to Copenhagen for a term to listen, learn and contribute. Mott wrote to his parents soon after his arrival:

It was quite thrilling at Bohr's on Saturday evening. Lots of people there: Mrs Bohr, and the two eldest children, handing round delicious eats and drinks. Pauli sitting on a chair and rocking his fat body about and telling humorous Jewish stories. Gamow talking about Russia. And Bohr going round and talking shop to one person privately after another; fearfully eager looking, asking about one's own work, glowing when he talks of the big problem yet to be solved, that is to be discussed when Heisenberg comes here. That is the problem, perhaps, of the interconnection between Relativity and the Quantum Theory, which contradict one another rather.[1]

Mott and Gamow, both tall and thin and neither able to speak the other's language properly, became boon companions. Mott was thrilled by the Russian's tales of war and revolution in Odessa, of the fighting he had witnessed between Reds and Whites, of occupation by Germans and French, of typhus and cholera and hunger. 'And so, Motti,' Gamow would say as each story came to its end, 'what a pity you were not there.'[2] They went to the cinema together, walked in the country together, played ping-pong together and most days Mott would lend Gamow money to buy cigarettes. Meanwhile Gamow was establishing himself as the court jester of the institute, always staging practical jokes, writing poems and songs and organizing parties where people played parlour games – a favourite involved everyone lying on their backs and passing balloons around with their feet. He was the soul of the party and people loved his talk all the more because of the breathtaking liberties he took with language. Although he spoke neither German nor English nor Danish with any accuracy he did not hesitate to hold forth in all of them and the letters he wrote in this period bear witness to the anarchy of his vocabulary and syntax. He seems, too, to have cultivated the impression that toil was foreign to him. Although the institute was open from 10 a.m. he made a habit of arriving late and colleagues learned that on his bedroom wall was the framed verse:

> When the morning rises red
> It is best to lie in bed.
> When the morning rises grey
> Sleep is still the better way.
> Beasts may rise betimes, but then
> They are beasts and we are men.[3]

The air of idleness, though, was probably misleading, as another line from a Mott letter suggests: 'He reads Conan Doyle and doesn't go to concerts, but is a brilliant physicist and hard working . . .'[4] In fact Gamow was still thinking about the nucleus, had some important ideas and was working – slowly, perhaps – on a new article for *Zeitschrift für Physik*. In the last week in September, however, news arrived which took his breath away.

It came in the weekly London journal *Nature*, where among the short communications appeared one from Ronald Gurney and Edward Condon of Princeton University entitled: 'Wave Mathematics and Radioactive Disintegration'. The title alone told the story but the eight paragraphs of text confirmed it with devastating clarity: Gurney and Condon had made exactly the same discovery as Gamow, and since his first article had still not been published they had beaten him into print. It was all there: they described the energy barrier around the nucleus and the two states of the alpha particle, inside and outside, and they noted that under classical laws the transition from one to the other was impossible. When Schrödinger's 'wave function' was applied to the problem, they showed, it could be seen that there was always 'a small but finite probability' that an alpha particle which had been inside might appear outside the barrier. They described the process by which this probability might be calculated and went on to say that their method satisfactorily resolved several outstanding physics problems. It was all very neat and tidy; so much so that they concluded:

Much has been written of the explosive violence with which the alpha particle is hurled from its place in the nucleus. But from the process pictured above, one would rather say that the alpha particle slips away almost unnoticed.[5]

This note to *Nature* was dated 30 July by the authors, one day later than the date Gamow had marked on the foot of the typescript he had sent to *Zeitschrift*, but that scarcely mattered now. Gamow's article, the work which had so impressed everyone in Copenhagen, risked becoming a mere footnote in physics history. That it did not he owed largely to the quick reflexes of Niels Bohr.

Bohr saw that Gamow must publish his latest ideas as quickly as possible, first to be sure that he was not anticipated again and second to make his mark as noticeably as possible in the developing debate. He could not wait for *Zeitschrift* this time, since they already had two of his papers in the works there and might not be able to bring out a third before the year's end. The obvious course was to publish a direct reply to the

Princeton article in the pages of *Nature*, which was a weekly and would move more quickly. Could they make this happen? They had some high cards to play. Bohr's standing would no doubt help convince *Nature* of the article's importance, but much better than that they had a direct line to the journal's editor. Visiting Copenhagen that week was Ralph Fowler, who had come from Cambridge with Eileen for a week or so to stay with the Bohrs while he joined in the discussions at the institute. Fowler thus heard at first hand from Gamow about the new theory of alpha escape, a theory he thought 'very beautiful', and he was also present to see the young Russian's reaction when the Princeton men stole his thunder. That both Fowler and his father-in-law were well known and respected at *Nature* may well have influenced the decision about what should be done next.

Gamow was quick to perform his part. By 29 September – in less than a week – he produced a short article for publication, complete with two diagrams. This was a condensed version of the paper he was writing for *Zeitschrift*, translated no doubt with assistance from Mott, Fowler or Bohr, or possibly (given his difficulties with language) all three. Since Fowler was leaving for England at that time he may well have carried the typescript personally to the *Nature* office in London, on his way to Cambridge. Certainly by 5 October he was able to write to Bohr: 'I will send Gamow's proof after correcting it myself to you for revision. I have warned the Editor [of *Nature*].'[6] Rutherford became involved, and made a couple of suggestions to improve the paper, so when the proof reached Gamow in Copenhagen the Russian had corrections to make before sending it back to London. Everything was done with the greatest dispatch, although even the combined weight of Rutherford, Fowler and Bohr could still not secure publication before 24 November.

As events unfolded, however, the date of publication was not the important one. Much more significant for the history of nuclear physics was the establishment, by Bohr and Fowler, of a direct link between George Gamow and the Cavendish Laboratory. It is likely that Bohr always intended to make this connection at some time but there is no sign that it was in Gamow's mind. It was the peculiar, urgent circumstance of the Princeton article, combined with the fortuitous presence of Fowler in Copenhagen, that precipitated matters, and the correspondence makes clear that the link was established in the first days of October 1928.

Once Fowler had shown Gamow's article to Rutherford there can be no doubt that word got out quickly among staff and students at the Cavendish. The Gurney–Condon paper must already have caused a considerable stir since its conclusions appeared to alter the theoretical basis of much of the work done in the laboratory. But Gamow's new article for *Nature* took things one step further, and it was a vital step, for instead of discussing how particles managed to get out of the atomic nucleus it looked at how they might get in. Essentially, he said, the same rules applied. By employing wave mechanics it was possible to calculate the probability that an alpha particle would enter the nucleus even when, on the classical theory, it did not have sufficient energy to climb over the protective barrier. This had been happening in the artificial disintegration experiments of Rutherford, Chadwick and others for nearly a decade; Gamow now supplied the means of explaining it. His findings conformed very neatly with Rutherford's past results, showing for example how, at a given energy, penetration became progressively less likely the heavier the element bombarded. In just this way, Rutherford's natural alpha particles had proved effective only against light elements. This was indeed a big step forward, but somebody quickly saw that it foreshadowed another even bigger one, and that person was John Cockcroft.

There was a side to Cockcroft that was so easily overlooked that even his close colleagues rarely mentioned it in their later reminiscences of him. He stood out at the Cavendish as a person who got things done and because his background in electrical engineering and industry enabled him to contribute so much to Kapitza's work. But Cockcroft was also something of a theoretician, indeed by the standards of the Cavendish researchers he was a highly qualified and able one. His first brief spell at university before the war had been devoted to mathematics and he had switched to engineering after his return to civilian life only because he thought it offered a better prospect of paid work. Later, when he left Metro-Vick for Cambridge, it had been to take a degree in maths, although he saw to it that the mathematics he studied were on the borderline with physics. This, as we have seen, was the nearest Cambridge offered to a training in theoretical physics and it marked out Cockcroft, soon after his arrival at the Cavendish, as a man peculiarly well equipped to follow the strange developments on the continent from 1925 onwards.

The precise sequence of events as 1928 drew to its close is impossible to establish because the written evidence is scanty and because the participants, when they looked back years later, gave accounts clouded by the passage of time. But we have one important clue. Among Cockcroft's collected papers at Churchill College in Cambridge is an article by Gamow – not the one he wrote for *Nature* and signed on 29 September, but the much longer paper he was writing in German for *Zeitschrift für Physik*. It is not an abstract from the published journal but a carbon copy of a typescript, with diagrams and Greek characters added in pencil, no doubt by Gamow's own hand. The final paper that went forward to the journal would carry the date 10 November, but this draft is clearly dated October. Cockcroft would later refer to it as an 'advance copy' and say that Gamow sent it to him from Copenhagen.[7] It was this that provided the Englishman with the detail and the information he needed to make the final step in this particular journey.

When Rutherford had made his speech to the Royal Society a year earlier, calling for high voltages to accelerate particles, it was a premise of his appeal that the voltages would have to be sufficiently high to surmount the atom's protective barrier. That was why he spoke in terms of 8 million volts and more, a daunting, even outrageous proposition at that time. But on seeing Gamow's latest paper Cockcroft realized that the Russian had changed the ground rules and he set to work with some calculations of his own. These also survive on paper, a short and dense piece of mathematical work whose gist is none the less simple. Using Gamow's formulae for the probability of an alpha particle entering the nucleus, Cockcroft estimated the likelihood of a *proton* doing so. Since protons carried half the electrical charge of alpha particles (+1 as against +2) it seemed that they would encounter less repulsion when they met the positively charged protective barrier of a nucleus, and so their chance of penetration should be higher. And that chance, it turned out, was not merely twice as high. Cockcroft was able to show that, on Gamow's equation, protons should in general be considerably more effective as disintegrating projectiles than alpha particles moving at similar speeds. Especially interesting was what happened as the energy element of the equation was reduced: very quickly the potency of alpha particles tailed away to negligible levels, but with protons the decline was much less sharp. Astonishingly, the calculations suggested that protons subjected to acceleration by as little as 300,000 volts continued to have a significant chance of penetrating the nucleus of a light element.

This transformed at a stroke the prospects for disintegration by electrical machinery. Instead of requiring 8 or 10 million volts for the task, just 300,000 might be sufficient – perhaps a single Tyrannosaurus Rex instead of six. To John Cockcroft's mind, one large carnivorous dinosaur would certainly be difficult company in the laboratory but it might just be manageable, and the game was worth the candle. He typed up his calculations on a single sheet of paper under the title 'The Probability of artificial Disintegration by protons' (this survives, with the same random capitals and numerous corrections throughout) and showed them to Rutherford.[8] Exactly when he did this we do not know because the sheet is undated, but it seems likely that it was in October or possibly early November. Of the import of the document, however, there is no doubt whatever: it amounted to a proposal for a completely new attack on the nucleus. Protons would be the projectiles and electrical power of around 300,000 volts would provide them with the necessary momentum. If it worked – if Gamow and Cockcroft were both right – Rutherford would no longer need to depend on natural radioactivity to break down nuclei; he would have a machine to do it, with a beam that he could turn up or down, on or off, at will. And from there he might move on to bigger machines, with more power, capable of cracking open any nucleus he chose.

The idea had grown in stages. It began in Göttingen, when Gamow saw that a wave equation could account for alpha particle escape. The next stage came in Copenhagen, when he put his thinking in reverse and calculated the probabilities for the penetration of nuclei by alpha particles. Then came the vital link with the Cavendish, the only laboratory truly dedicated to the experimental investigation of the nucleus,

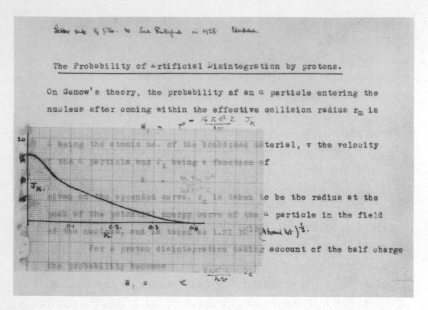

Cockcroft's memo to Rutherford, with graph attached

and that brought the final stage, as Cockcroft took Gamow's calculations and adapted them for protons. The whole process included chance occurrences, delays and apparent setbacks but from start to finish it needed just three or four months. Now the question was, would Rutherford believe it, or rather, would he believe it sufficiently to allow Cockcroft to put the theory to the test?

It is possible, sifting through the writings and sayings of Ernest Rutherford, to form the impression that he had an uncomfortable, even a resentful relationship with the world of theoretical physics. His own discoveries were always founded four square on laboratory work, if not his own then that of his close collaborators and subordinates. This is not to say that he did not think for he certainly did, and profoundly, and the interpretations he placed upon his laboratory findings reflected a supple and ingenious mind. But his heart was in the laboratory and he would have been appalled if anyone had referred to him as a theoretician. To him they were a different breed – it is likely that on some visceral level he could not understand how, if they were genuinely curious about atoms, they could resist rolling up their sleeves and doing some experiments – and in his brasher moments he was not above teasing them. He once remarked that the theorists 'play games with their symbols, but we, in the Cavendish, turn out the real solid facts of Nature'.[9] Of Einstein's theories he was said to have declared: 'That stuff! We never bother with that in our work.'[10] And at the height of the excitement over quantum mechanics he pronounced grumpily that the theorists were 'up on their hind legs again and it is time we shot them down'.[11]

It is Rutherford's actions rather than his words that tell the true story. Some of his best friends were theoreticians, the most striking example being Bohr. The friendship begun in Manchester was sustained for the rest of his life through visits and correspondence, and that correspondence is striking for the depth of mutual respect it shows; both men recognized the interdependence of theory and experiment. In the context of British academia between the wars Rutherford may even be described as a champion of theory, for at Manchester he created a chair of what was called mathematical physics (in the vain hope of luring Bohr back), and he was to do the same somewhat later in Cambridge. And then there was Fowler, Rutherford's son-in-law and a genuine intimate. Not himself a ground-breaking thinker by international standards, Fowler was an inspiring teacher and a man of great dynamism who took on himself the task of keeping the Cavendish professor abreast of development in theory.

Rutherford thus had respect for theoretical physics even if it was not a discipline to which he liked to apply his own mind. As for the new quantum mechanics, in the very month when Cockcroft was cooking up his idea about proton acceleration the professor gave his views on this in public. A guest speaker at the annual dinner of the Institution of Mechanical Engineers, he was introduced with a joshing remark about quantum theory and replied: 'May I say, in order to remove all doubts, that the older form of the quantum theory [the Planck version] is completely dead, but another and more terrifying theory has taken its place.'[12] This newcomer was the wave theory,

in which we regard a particle not only as a particle, but also as a wave. There is a sort of duality of things, though whether the conceptions underlying the wave theory are purely mathematical or correspond to reality we do not know. We do know that this new avenue of attack is proving surprisingly successful in the directions in which it can be applied.[13]

These sentences express beautifully the fine balance in Rutherford's mind at the time when he came to consider the implications of Gamow's work. It was characteristic of him to be unsatisfied with the 'purely mathematical' and to wish to see it tested against 'reality', but it is equally plain that he accepted the importance of the ideas, however 'terrifying' they might appear, and was prepared to acknowledge their successes. When Cockcroft went to him and presented his ideas, therefore, the door was ajar; Rutherford was open to the suggestion that quantum mechanics might help the Cavendish. And so, when he was told that it offered him the possibility, if no more than that, of escaping from his depressing experimental morass, his interest was aroused. He told Cockcroft to start work.

Since he shared a workroom with Allibone and Walton and

they were the people in the Cavendish most concerned with accelerating particles, Cockcroft's first step was naturally to discuss the idea with them, and while Allibone had plenty of advice to offer, Walton was able to give some practical assistance. Having just commissioned the new version of his linear apparatus from the workshops he had some time on his hands while he awaited delivery and so he joined Cockcroft in some preliminary experiments and planning during that November and December. A 300,000-volt accelerator would be a large and complex machine, vastly more so than Walton's own device; there was a great deal to be done. The main steps involved seemed clear. First would come the power supply – the force that was to push the protons – and this would require specialized machinery. Then there would be the glass acceleration tube – the vessel along which the protons would be pushed – which must be designed in such a way that it could tolerate the high voltage. Third would come a source of protons, both a device for producing them and a means of introducing them into the acceleration tube. Last would be the means of putting the fast particles to use in experiment and observing the results. It might be 300,000 volts rather than 10 million but each of these stages was beyond the state of the art, even at the Cavendish. Considerable thought and experimentation would be needed in the design phase alone, indeed so much that it was difficult to know where to begin, but with Walton's willing help Cockcroft made a start.

In January George Gamow came to visit. Bohr suggested the trip and Fowler issued the invitation, but predictably Gamow complicated things by deciding to travel via Hamburg, to call on the theoretician Pascual Jordan, and Leiden, to see Ehrenfest. All went well in Hamburg but when he reached the Dutch border he was refused entry – he was a Soviet citizen and possibly a dangerous communist, and he did not have the appropriate visa. It took the personal intervention of Ehrenfest, a figure of standing in the Netherlands, to get him across the frontier, although when they met the two men do not seem to have hit it off. One reason was that Ehrenfest was very busy and another was his loathing of the smell of cigarettes; Gamow's later memories of the visit would be dominated by the struggle to conceal his habit. From Rotterdam he sailed to England, eventually reaching Cambridge on 15 January 1929. He stayed three weeks, addressed a colloquium on his new ideas, met most of the leading lights of the Cavendish and conceived a deep affection for the place. A delighted Mott wrote afterwards, in a flurry of exclamation marks:

It was amusing having Gamow here! At first he thought Cambridge terribly highbrow, no *Dummheit Machen*, only terrible parties where one sits around and talks about one's soul! But together with the Hartrees we arranged some for him with balloons and brushes! Perhaps he has forgiven us.[14]

More importantly, Gamow impressed Rutherford and he extended his stay by a week when Rutherford invited him to contribute to a Royal Society discussion on the nucleus – a considerable honour. The record indicates that he spoke only briefly and in quite technical terms and it was left to Fowler to give a full presentation of the Gamow theory of alpha particle escape, which he did in memorable style. Patiently explaining that a wave could be a particle and a particle a wave, Fowler showed how these dual creatures were not confined by barriers in the familiar way:

They can steal through them – only rather occasionally, of course. You may say that any one of us present has a finite chance of leaving this room without opening the door, or, of course, without being thrown out of the window.[15]

Gamow's visit made a practical contribution to our story by further convincing Rutherford of the value of Cockcroft's proposal, and so by the time the Russian left in early February the laboratory director had made a decision which reflected a deepening commitment to the project. It came about after Walton suffered another setback in his research with the appearance of an article in a German journal, *Archiv für Electrotechnik*. Written by Rolf Widerøe, a Norwegian physicist and engineer, the paper described efforts made over several years to develop particle accelerators of exactly the kinds that Walton had been working on. Although Widerøe had not made the electron-spinner function he had had more success with the linear apparatus and, while there were areas where the Irishman could claim to be in advance of Widerøe's thinking, there was no doubt that the Norwegian had fuller knowledge and longer experience. More than that, he was in print, and even worse for Walton was the discovery from the footnotes that he had also been duplicating other work published five years before by a Swede, Gustav Ising. It was a serious blow, though Walton insisted later that he took it philosophically: 'No injustice had been done to me and so I had no cause to nurse any grievance or ill feeling. As far as I can recall I merely decided to try to be quicker off the mark next time.'[16] Fortunately he was not left to dwell on it, for Rutherford immediately gave him something else to do. He was to join Cockcroft full-time in developing the 300,000-volt accelerator. From the director's point of view this was the obvious course because Cockcroft was not well placed to put his own idea into action. He already had plenty of work on his hands around the Cavendish and was not in any case the laboratory's most gifted experimenter. If proton acceleration was really to be put to the test, therefore, he would need a partner who could give it his full attention and was good with apparatus, and Rutherford was sufficiently impressed by the young Irishman to give the job to him. From that moment on, therefore, it was Cockcroft and Walton.

7. Hardware

If Cambridge had been timorous about the new railways in the middle of the nineteenth century it showed more courage when electricity came along a few decades later. A coal-fired power station opened in 1892 – early by national standards – on a site just a short walk from the colleges in the north of the town. Employing the most modern steam turbines of the time, it offered customers a 100-volt supply along lead-covered cables, although this was for use in lighting only and no power at all was produced in daylight hours. Over the following thirty years the service improved but progress was slower than we might imagine: while daytime supplies had arrived by the 1920s and electric irons, kettles and wirelesses were appearing in the better-off Cambridge homes, it is striking that in Walton's first lodgings the only electric fitting was still the light – and for that he had to pay extra. In fact, across Britain as a whole in 1927 most houses were still lit by gas and in the countryside electricity was barely known. As for the Cavendish Laboratory, it made no use of the mains supply in experimental work; it was not reliable enough to keep a clock running to time so it could hardly be trusted in research. Instead the lab had a motor generator of its own, located behind the workshops and providing current for the workrooms at 200 volts, which was better suited to the researchers' needs. Anyone who required something more powerful was issued with a long wooden box like a coffin, which contained several generators in series and gave an output of 2,000 volts.

These primitive arrangements in both town and laboratory

were soon to be superseded, for in Britain as in America the electricity industry was at last on the march. The construction of the British National Grid began in 1927 and the aim was to complete it, working from north to south, within five years, thus assuring more efficient and general coverage as well as cheaper supplies to customers. In anticipation of these changes Cambridge's local power company was already merging with a larger group and over the next few years the service in the town would be improved to meet national standards, and in due course linked to the grid. At the Cavendish this meant not only that the noisy generator could eventually be turned off but also, incidentally, that the British power industry was acquiring some useful expertise. As Rutherford pointed out in his Royal Society address, high voltages were the most efficient means of transmitting power over long distances. Power stations, often located for convenience in coalfields, usually produced a supply at around 15,000 volts, but to transmit this over the grid to consumers in the cities would have entailed unacceptable losses through heat and resistance. Instead the supply was effectively forced along the cables under pressure, the voltage being 'stepped up' to 132,000 volts for transmission and 'stepped down' again at the other end before being distributed to homes and businesses.

For Cockcroft and Walton the first and in many ways the simplest requirement in building their apparatus – it was simple for them, at least – was a source of high voltage. They needed an up-to-date motor generator and they needed one of those machines used to step up voltage, a transformer, and for both the obvious place to go was Metropolitan-Vickers, Cockcroft's old firm and one of the principal suppliers to the grid. At Metro-Vick's Manchester works their request for a generator raised no difficulties at all but the transformer was a different matter. High-voltage transformers were made for open sites and factory spaces and were rarely subject to constraints of size; even the ones used in hospitals to power X-ray equipment were as big as motor cars. The two scientists, however, needed a machine that would not only pass through the Cavendish archway but could also be manoeuvred down a flight of stairs, along a corridor and through the doors into their workroom. And once there it had to leave enough space both for them to conduct their experiments and for Allibone to continue his. None of the models in production at Metro-Vick was small enough to do that and still provide the required voltage. Cockcroft discussed this problem at an early stage with his friends at the company labs, notably Brian Goodlet. One of several very able physicists there, Goodlet was a British subject who had been brought up in St Petersburg and was reputed to have escaped the Russian Revolution by shooting his way down the Nevsky Prospekt. He was friendly not only with Cockcroft and Allibone, both former colleagues, but also on more formal terms with Rutherford, and having provided equipment for the Cambridge laboratory in the past he was already familiar with the requirement for compactness. 'That man,' he wrote irritably of the Cavendish director, 'ought to be made to look at things through the wrong end of a telescope, since everything is too big for him.'[1] How great were the difficulties, Goodlet was now asked, in making a wardrobe-sized transformer capable of giving 350,000 volts, and how much would it cost?

Transformers are relatively simple and robust devices which raise voltage in something like the way that squeezing the end of a garden hose produces a more powerful jet of water – the pressure (voltage) is increased while the volume (current) is constricted or reduced. The art of making them was largely a matter of winding coils of wire, arranging them suitably and ensuring that each part was insulated. Goodlet knew, however,

that squeezing a 350,000-volt capability into a box the size of
a wardrobe would demand extremely fine wire, very tightly
wound and laid out in novel ways, which in turn meant
developing new production techniques. He calculated that
the job would take six months or more, roughly until the
autumn of the year. To the Cambridge physicists, who had
more than enough other work on their hands, this was an
acceptable delivery date, and the price must also have been
acceptable although we have been left with only a rough
idea of what it was. At about this time Cockcroft persuaded
Rutherford to apply to the university authorities for a special
grant of £500 to support the project, and the payment was
approved. It is likely that the bulk of this sum (almost double
Walton's annual income) was spent on the transformer.

Next the two scientists turned their attention to the rest of
the apparatus. This, as we have seen, would include the proton
supply which would provide the projectiles for the accelerator
as well as the glass tube in which those projectiles would be
channelled. And by now it was clear that another important
component would have to be added: rectifiers. Over the life
of the project, as things turned out, it was these that were to
cause the greatest heartache. The need arose because Cockcroft and Walton saw that it was not enough merely to have a
supply of 300,000 volts (or 350,000, as they had specified to
Metro-Vick to be on the safe side); it also had to be the
right *kind* of electricity: d.c. rather than a.c. By their nature
transformers produce only a.c., or alternating current, which
constantly switches direction, travelling clockwise in a circuit
at one moment and anticlockwise the next. Almost all our
electrical appliances today operate on this type of current
although we do not notice it because the reversals of direction
are so frequent (typically fifty or sixty per second) – the
filament of a domestic light bulb in an a.c. circuit, for example,

does not have time to dim visibly between the alternations. What the two Cavendish scientists were aiming to do, however, was not to keep a simple filament aglow but to propel protons away from a very powerful positive electrical terminal down a glass tube and then use them in experiments, and it was clear that for this purpose a current that constantly switched directions would not be suitable. Instead they needed one-way or direct current (d.c.), and for this they would have to 'rectify' the a.c. emerging from their transformer. A special device to do this job would have to be fitted between the transformer and the rest of the apparatus, and since no commercial rectifiers of the right capacity were available the two scientists were obliged to design and make their own.

A rectifier works a little like a weir in a tidal part of a river: the river water can flow down over the weir without interruption but an incoming tide is halted and can go no further upstream. Electrical current is a flow of electrons pushed by a battery or other power source and in alternating current, as we have seen, these electrons rush first this way and then that. They flow easily along a conducting material such as a loop of copper wire but if you cut the wire that naturally changes: the circuit is broken in both directions. With a rectifier you can restore the connection in such a way that the current will flow in only one direction. You make one by feeding the two tips of the cut wire into opposite ends of a glass vessel and then pumping the air out of the vessel. Glass is an insulator and so does not conduct electricity and current will not normally cross a vacuum either, so the circuit remains broken. If, however, you then apply heat to one of the wire tips inside the glass a current will immediately start to flow from that hot wire along the length of the vessel to the other wire. This happens because the act of heating the wire liberates electrons in the metal and, pushed by the power source, these are able to race across the vacuum to the other wire, carrying the current and completing the circuit. This effect was very well known by 1929 and what made it useful to Cockcroft and Walton was that the current would flow *only* from the heated wire to the cold one. It could not go in the other direction because electrons would not leave a cold wire. In an alternating current circuit, therefore, the 'forth' element would flow freely while the 'back', blocked at the cold wire, would be eliminated. This was the electrical weir.

There were daunting difficulties when it came to applying this principle in a high-voltage assembly so to make the job more manageable the two physicists decided to build two smaller rectifiers rather than one very large one. Even so it was a challenge to design something that would cope with the potentially violent qualities of a 300,000-volt supply. The glass tubes (scientists always refer to glass vessels as tubes) would have to both contain and withstand the effects of something which, if it escaped its bonds, was capable of leaping suddenly and unpredictably through the air to make contact with the nearest earthed object – effectively a flash of lightning in the laboratory. In Allibone's experiment with a rat a spark of this kind had drilled a hole half an inch in diameter through the unfortunate animal and there was no doubt that it would be equally deadly to a man. And even when the tubes did everything that was asked of them there would be another hazard to reckon with: corona. In the corona effect electrons were drawn to sharp points, edges and even specks of dust on the apparatus, effectively rendering it 'live'. Sometimes corona produced a miniature electrical storm on the surface of the glass, with twitching blue lights and a sinister hiss. And it could be destructive too. In early experiments, when Walton borrowed Allibone's Tesla apparatus and subjected a glass vessel to high voltages, he was able to watch these 'surface creepage sparks' dart around the outside of the tube and then, on finding a point slightly weaker than the rest, punch clean through the glass, instantly destroying his vacuum. Corona was to prove a subtle and persistent adversary.

The size and shape of the tubes was crucial. There was a known relationship between the length of a tube and the voltage it could be expected to withstand before corona took over, and for Cockcroft and Walton's purposes the necessary length was about one metre for each rectifier. Shape was important because rounded forms resisted corona most effectively and it was probably because of the advocacy of Allibone that their choice fell on what came to be called 'bulbs'. These were tubes of glass the required metre in length, like a drainpipe at each end but bulging in the middle third into a shape similar to a rugby ball, though rather larger. Tubes of roughly this kind were used in X-ray machines in hospitals and since Allibone employed something similar on his own electron-acceleration machine he could show that it would cope with the stresses involved. It was, however, far too large an object for the scientists to make themselves and too large even for the Cavendish's professional glassblower, Felix Niedergesass; this would be another job for industry. But before they could fix the dimensions for their bulbs the two physicists needed to know more about what was to happen inside them.

Instead of plain severed wire inside the rectifiers there would be custom-made metal terminals. The heated one, which was the negative terminal, was known as the cathode, while the cold, positive terminal was called the anode, and together these were the electrodes. They had to be designed in a way that would allow a powerful voltage to pass from one to the other while at the same time minimizing the risk of corona. Cockcroft and Walton at first tried an arrangement in

which the hot cathode, which looked like a light–bulb filament, was almost enclosed in an anode shaped like a tin can
open at one end. The idea was that the electrons could flow
freely from cathode to anode and yet their chances of escaping
towards the glass and contributing to corona damage would
be greatly reduced. Unfortunately testing showed that this
caused an unacceptable drop in voltage and so the two electrodes had to be moved apart, with a distance of about five
centimetres giving the best results. The detailed design of the
little cathode itself presented some difficulties and it was only
after a lot of hard work that they settled on a 'V' shape made
of tungsten wire. Next, of course, this filament had to be
heated independently so that it would release electrons. It was
all very fiddly and there was no question of simply buying
parts ready-made; everything had to be tailored to the needs
of the machine.

For the rectifiers alone, in other words, numerous components had to be drawn, made, tested, measured and understood, and a good many of the designs depended upon one
another – alter one and you could find that all the others had
to be altered too. Through the spring and summer of 1929
Cockcroft and Walton constructed a series of experimental
rectifiers and tested them using Allibone's formidable high-
voltage apparatus, when it was available (Cockcroft would
write wryly of 'considerable congestion around the Tesla
set'[2]). Trial and error were followed by more trial and more
error until at last a set of workable compromises was found.
Only when that moment came were they able to settle the
specification for the glass tubes: each would have a stem of nine
centimetres diameter at one end and one of five centimetres at
the other while in the middle the bulb, very carefully and
evenly shaped, would widen to a maximum of just under
thirty centimetres. After consultation with specialists it was

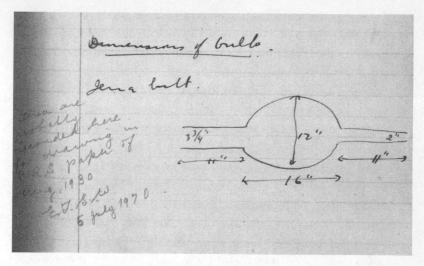

From Walton's notebook: an early sketch of a rectifier bulb

decided that the glass must be made with the metal molybdenum, an element known for its hardening properties. Outside the tubes they would set up 'corona shields', metal parts
designed to cover sharp edges and draw stray electrical charge
away from the tubes, preventing punctures. The tubes themselves were ordered from a German company in Jena and it
seems, on the evidence of a remark in a Cockcroft letter of
the time, that they got a good price.

The scientists also had to find the best way of drawing the
air out of the rectifier tubes and here they had a remarkable
piece of good fortune, for they became the first researchers
anywhere to take advantage of an important advance in technology. Several commercial vacuum pumps were in use in
laboratories in the 1920s but none had been found that could
be said to make the physicist's life easy, or which was cheap
enough to suit a university budget. Typically two separate sets
of equipment were used, the first a simple mechanical pump

to suck out the majority of gas and the second a device which trapped or washed away as many as possible of the stray molecules that remained. In 1929 a 'mercury diffusion pump' was generally preferred for the second and more difficult task, but these were expensive and the Cavendish could afford only three or four for the whole lab. Researchers usually had to make do with a much older system that relied on the ability of charcoal to absorb gas, and this could be heartbreakingly slow. Whichever system was used, however, there was a further, very troublesome complication: they both required regular topping up with liquid air. For this purpose the Cavendish maintained a large, cantankerous machine (a gift, appropriately, from the Air Ministry) which cooled the cocktail of gases we call air to -187 Celsius, the point at which it became liquid, and each Monday the scientists would have to report with their flasks for a fresh supply. The machine was costly to run and dispensed only grudging quantities so economy was encouraged. Among Rutherford's most intimidating remarks to a research student, made at the height of one of his rages, was: 'The time has come to consider whether your experiments are worth the liquid air they consume!'[3]

Cockcroft and Walton, who would have to evacuate vessels far larger than any others in use in the laboratory, were not only spared the need to use the old charcoal equipment but they also escaped entirely the requirement for liquid air. In fact they were able to use technology more advanced and more easily managed even than the top-of-the-range mercury pumps, and for this they had the Metropolitan-Vickers connection to thank. In 1927 a Metro-Vick scientist called Bill Burch had developed a new way of distilling oil which yielded several products with interesting properties, among them one which, he discovered, could substitute for mercury in a vacuum pump. The oil, which he called Apiezon, was much cheaper and easier to handle than mercury but best of all it did away altogether with the need for liquid air. By 1929 Metro-Vick was half-way towards the commercial production of Apiezon pumps and John Cockcroft, getting wind of this development, was quick to snap up a couple of prototypes at the bargain price of fifty shillings apiece. Well before this revolutionary labour-saving item came on the market, therefore, he and Walton were able to use it in their apparatus and it is worth remembering that, though they were to endure many frustrations in the years to come, they were extremely fortunate never to have to struggle with the old-style pumps or hustle for liquid air. And this was not the last favour they would owe to Bill Burch.

The new pump had of course to be connected to the rectifier tubes so that it could extract their contents and this was no routine matter of inserting a pipe because the high voltage would not permit it. The pump connection would have to be insulated, a task eventually accomplished by the cumbersome means of adding to the apparatus a third big glass tube, similar to the others. Things by now were becoming very complicated, with the anode, the cathode, the heating for the cathode, the bulbs, the pumps, the evacuation tube, the corona shields and other components – and all of this was still only what was needed to make the rectifiers work. Cockcroft and Walton were busy at the same time testing possible proton sources to provide their projectiles and progress on that was equally deliberate, while they had yet to begin planning their acceleration tube. The autumn term of 1929, therefore, still found them months away from even attempting to accelerate particles. If this pace seems slow the difficulty of the work was largely to blame, but it is also true that the Cavendish Laboratory in those years was not an institution built for speed.

8. Lab Life

At Cambridge there were three university terms for the under-graduates, Michaelmas, Lent and Easter, each lasting ten or eleven weeks, while the research workers at the Cavendish had a fourth, called the Long Vacation Term, during the summer. That gave a total of about forty weeks in the year during which everybody was expected to attend the laboratory and carry out experiments. If you wanted to work beyond those terms you might be allowed to squeeze in an extra few days here and there but for at least eight weeks of the year the place was simply closed – the generator was turned off, the great oak doors to Free School Lane were locked and that was that. In these vacations scientists were expected to rest, take holidays and do some reading. Younger students from distant dominions often embarked on tours of Britain and Europe but Walton, no great traveller, tended to go home to his family in Ireland. In younger days Cockcroft had enjoyed climbing in the Alps but now he preferred touring holidays and seaside trips. The Rutherfords, for their part, usually adjourned to a cottage they owned near Betws-y-coed in North Wales. Sometimes groups from the lab made educational trips to laboratories in Germany or France, and September brought the season for scientific conferences – Cockcroft attended some of these but there is no record of Walton doing so. Whatever they did it is clear that, no matter how important or delicately poised the experiments in progress, work at the Cavendish was subject to frequent interruption by holidays: a couple of weeks at Christmas, again at

Easter and again in early summer, and perhaps four weeks in late summer.

The daily timetable could also inhibit progress. Workshop technicians and other support staff arrived earlier but for the research scientists the day usually began gently at around 10 a.m. and even then no one paid particular attention to the clock. There was great rigour, by contrast, about the end of the day, which always and for everyone came at 6 p.m. sharp. The rule had been kept since the days of J. J. Thomson: whatever the state of your experiment, even if you were on the brink of completing some vital task, you had to stop what you were doing and go. A technician would progress from workroom to workroom when the hour had struck pulling out all the plugs and throwing all the switches, although Chadwick would later insist that this was not so brutal as it seemed. By his account it was the head of the workshops, Fred Lincoln, who roamed the corridors calling 'Time to go' and it was only for safety's sake that he ensured everything was switched off. Even Chadwick admitted, however, that Lincoln was not to be argued with, and told the story of a young New Zealander who was still winding the handle of an old vacuum pump at 6 p.m. On being told it was time to go he asked merrily, 'Can I play a tune for you, Mr Lincoln?' as if he were operating a barrel organ, and received the brisk reply: 'Yes, *'ome Sweet 'ome*, if you don't mind, Mr Burbidge.'[1] When you consider that many of these students were working with vacua that would inevitably leak and with radioactive sources whose effectiveness declined by the hour it is easy to imagine how frustrating the deadline could be.

Rutherford, who in his younger days had often burned the midnight oil, not only accepted but wholeheartedly endorsed his predecessor's rule, declaring that he had seen researchers ruin their health through overwork and he would not have that happening in his lab. He also believed with some passion that experimental work was worthless if it was not accompanied by reflection and analysis, and from his own experience he felt this could best be done at home. When research students begged him to let them work late he would always tell them firmly to go home and do some thinking. Visiting scientists had long viewed this arrangement with astonishment. A letter survives, written by an American physicist to J. J. Thomson, declaring: 'A laboratory in this country in which nobody ever began work before 10 a.m. or worked later than six in the evening would serve as a terrible example of sloth and indolence.'[2] The same correspondent, it is only fair to add, admitted that despite living so 'pleasantly and unhurriedly' the Cavendish staff seemed to be remarkably productive. And they did work on Saturday mornings, although so did most of Britain in those days.

There is no record of Cockcroft or Walton complaining about this regime, which they observed faithfully, but the six o'clock rule did stir up some feeling at Metropolitan-Vickers when the time finally came for the delivery of the transformer. Although the machine had been promised by the middle of October 1929 there was a delay whose cause, considering that this was the month of the Wall Street Crash, is ironic. To complete its insulation the transformer was to rest in a tank of oil, and Metro-Vick relied on a specialist contractor to supply these tanks. When the moment came to place the order, however, it was found that the contractor's books were full and they were working to capacity, so delivery of the finished transformer to the Cavendish had to be put back to when the tank would be ready in December. As that date approached, Cockcroft's friend George McKerrow, who managed Metro-Vick's liaison with the academic world, wrote to say that – free of charge – he was sending two technicians to Cambridge to handle the installation. He went on: 'Please will you arrange for Smethurst and the wireman to be able to work overtime in order to get the wiring and general fitting up finished as rapidly as possible . . . It is, as you will realize, quite ridiculous that it should be shut up and everybody have to leave at 6 p.m.'[3] Since the job was done within the week it seems likely that for once Rutherford consented.

Between the hours of 10 a.m. and 6 p.m. the Cavendish scientists no doubt worked hard, although there were still interruptions to be reckoned with. As often as he could, which was perhaps once a fortnight, Rutherford did the rounds of the laboratory from 11 a.m. onwards and as we have seen that could sometimes be a disruptive, not to say frightening business – one student recalled scampering up a ladder to escape when he heard that angry, booming voice approaching down the corridor. When he was on good form, however, Rutherford was always a welcome visitor. He would bustle in, pull up a chair, a stool or if necessary a soap box and sit down. The pipe would come out and he would invariably beg a light, often pocketing the box of matches once the dry tobacco he preferred was aflame, and then he would inquire about progress. This was when he did most of his teaching. He was a thoughtful, encouraging supervisor and had an enormous reservoir of knowledge on which to draw when problems or mysteries arose. Occasionally, if a calculation were required, he would fish a stubby pencil from his pocket, seize a handy scrap of paper and scribble it out himself. And once the business was complete he would often linger, chatting happily. He might inquire about a student's lodgings or pastimes or holiday plans and from there the conversation would drift into pleasant reminiscence and anecdote – there were few physicists of note in the world whom Rutherford did not know personally and still fewer about whom he could not tell some revealing or mischievous story. Eventually he would catch himself, announce with a sudden irritation that he must get on, and make for the door, throwing over his shoulder a final exhortation to 'get on with the job' or 'cut out the frills'.[4]

Lunch for the scientists seems to have been a businesslike affair, though the lab had no canteen and it had to be taken elsewhere, but at 4 p.m. there was another Cavendish ritual: tea. In Thomson's day afternoon tea and cakes had been supplied and paid for by the director's wife (although she did not herself attend to pour) and Lady Rutherford kept up this tradition for some time. When the research staff grew beyond twenty-five or so, however, she drew the line. The packets of tea she continued to buy and deliver, but the cakes had henceforth to be purchased by their consumers, who set up a 'bun fund' for the purpose. In mid-afternoon, therefore, they would gather around the teapot in a little room beside the library to refresh themselves, exchange gossip, talk science and eat their penny buns. Rutherford occasionally appeared and would sit, but for the most part they had to stand, which may have been a way of ensuring they did not neglect their workbenches too long.

Another distraction, albeit a useful one, was meetings, which came in a variety of forms. Every second Wednesday afternoon the Cavendish Physical Society gathered in the main lecture hall, usually with Rutherford himself in the chair. This was the formal embodiment of the laboratory's research community and so the first meeting of every year, in the autumn, was like an AGM, with the director providing a *tour d'horizon* of progress both 'in house' and internationally. At other sessions there would be guest speakers or lectures from Cavendish researchers – anyone with an important paper coming up in the journals was likely to be asked to address this forum. Rutherford's introductions and closing remarks were often to be savoured, since when he disagreed with something, as he frequently did, he had the greatest difficulty concealing it. Nor could he conceal it when he fell asleep during lectures, which to his students' amusement he occasionally did. The audience, meanwhile, sat in the tiered benches where in past years students had heard Clerk Maxwell and Rayleigh, and at the half-way point descended to the speaker's level to drink tea. Every other Wednesday afternoon, when the Physical Society did not meet, Chadwick and Fowler convened their Colloquium, a forum whose job was to close the gap between theoreticians and experimenters. This was deliberately held in a room 'small enough to be full' so that discussion would be freer, and most of the papers were delivered by younger researchers on work in which they were not personally involved.[5]

Another fortnightly event was the evening meeting of the Cambridge Philosophical Society, which Rutherford did much to revive in the 1920s, and besides these official events there were several clubs, of which the most active was the Kapitza Club. Founded by the Russian soon after his arrival, this was restricted to a membership of about a dozen because they usually met in his college rooms. His idea was to foster a lively atmosphere of the kind he had known in his days in Petrograd, and that Gamow and his Jazz Club friends experienced there a few years later. Cockcroft and Walton were both members. Usually one person read a paper and then defended it against the other members but occasionally there was a quickfire procedure called 'five minutes' in which everyone in turn spoke for that period on some fresh topic.

When Cockcroft started at the Cavendish all of this surprised him, and he wrote after his first few weeks: 'Life is a round of meetings.'[6] In so small a community it must have been difficult to keep up attendances let alone maintain a flow of worthwhile papers, but it helped greatly that the lab had a steady stream of illustrious overseas visitors, often friends of Rutherford or J. J. Thomson. In Walton's first term, for example, the mathematician and politician Paul Painlevé was awarded an honorary degree at Cambridge, and gave a lecture to mark the occasion – followed by a reception where everyone was required to speak French. It was the first of many such occasions for the Irishman, who only a few years later could write: 'I think I must have heard or seen at least half of the most distinguished physicists of the world.'[7] And it was not essential for the visitor to be distinguished. In 1928 a budding German theoretician, Rudolf Peierls, was on holiday in England and casually called on a Cambridge acquaintance, only to find himself immediately roped in to lecture at the Kapitza Club. 'We have no speaker for this week's meeting,' he was informed, 'so you can talk there.'[8] He was only twenty-one and in his own words a raw beginner, but his hosts did not care: Peierls worked in Leipzig alongside Felix Bloch, so he must be able to give some account of what Bloch was up to. The occasion passed off satisfactorily.

Most evenings, after their expulsion from the laboratory, the research students dined in their colleges and afterwards retreated to lodgings if not to think, as Rutherford wished, then at least to read and write. The number of physics journals to be followed was growing and new books were also appearing almost by the week, especially in the field of theory. For a student such as Walton, who had arrived in the atomic world somewhat abruptly in late 1927, it seemed impossible ever to catch up with the necessary reading. Happily he was not required to buy all he read; the Cavendish library was kept up to date and graduates were allowed to take books home. Walton also did a good deal of thinking, calculating and writing at home. It was in his lodgings in Park Parade, on a typewriter which had been his first significant purchase on arrival in England, that he wrote the final draft of his M.Sc. thesis (which won him a cash prize and a medal from his old university). There too he prepared his annual reports to the 1851 Commission and his first published paper in atomic physics, a theoretical account of his attempts at electron acceleration for the *Proceedings of the Cambridge Philosophical Society*. Often he would bring home raw experimental results and work late checking and tabulating them, so that he would be ready to take the experiments forward in the morning.

Though he was certainly studious – the characteristic had been noted by friends at school who rearranged his initials, E.T.S.W., to make the nickname 'Stew' – Walton was no loner. Allibone and Rutherford were both struck by his friendliness and warmth and he relished the fellowship of college and laboratory life, in particular events such as the annual Cavendish dinner. These were raucous affairs held close to Christmas, at which ribald toasts were proposed and irreverent songs sung about the personalities of the lab. Peter Kapitza described one such evening:

You could do anything you liked at the table – squeal, yell, and so on – so the general picture was rather wild and quite unique. After the toasts everyone stood on their chairs, crossed arms and sang a song recalling old friends and so on. It was very funny to see such world famous luminaries as J. J. Thomson and Rutherford standing on their chairs and singing at the top of their voices. Finally we sang *God Save the King* and at midnight we broke up.[9]

Teetotal as he was, Walton joined in with gusto every year and was even known to perform verses of his own.

Like every other scientist in the lab, roughly once a term Walton had to brace himself for what was known as a 'royal command' – an invitation to Sunday afternoon tea with the Rutherfords. Well-meant though these occasions undoubtedly were, they could be anxious affairs for the younger scientists. Half a dozen guests would be present at Newnham Cottage, drawn from the professor's wide circle of acquaintance, and the lady of the house presided. Teacups and cake circulated while Rutherford kept up a flow of conversation, but only the couple's most seasoned friends found it possible to relax. Tense and birdlike, Lady Mary had the knack of keeping people on their best manners, not least because of her famously sharp tongue. (When her husband's portrait was unveiled at the Royal Society, for example, she is said to have expressed surprise that it showed him with his mouth shut.) She addressed every man as 'Mister', whatever his station in life, and in fair weather would often interrupt proceedings to lead everyone on a forced march around her garden. There was no risk, either, of anyone outstaying their welcome, for when she decided the moment had come she would briskly tour the room shaking everyone's hand in an unambiguous act of dismissal.

Of course it was possible in the evenings and at weekends for the scientists to escape from science. The cinema was popular (and just discovering sound) and there was the full range of university societies. Cambridge at this time was becoming a scene of left-wing political activity. Only a few years earlier many students had happily manned essential services to help break the General Strike, but things were changing and the economic crisis after 1929 would tip many towards socialism and communism. (Anthony Blunt was already around, a young postgraduate at Trinity, and Guy Burgess arrived in 1930, although their famous little group would not take shape for another couple of years.) Rutherford had no taste for politics of any kind but his subordinates at the Cavendish were touched by the new mood. One member of staff, Alec Wood, was a vigorous Labour Party activist and stood several times for Parliament, though without success, while another, J. D. Bernal, ranked among the university's most prominent Marxists. Patrick Blackett, meanwhile, was an enthusiast for all things Soviet who liked to invite colleagues home to watch films glorifying the collectivization of agriculture and the industrial achievements of communism, while Oliphant was a milder socialist of the *New Statesman* sort. However no one working in the basement room that was beginning to be known as the 'high-voltage lab' was attracted by left-wing ideas or for that matter was involved in politics of any other flavour.

In his spare time Allibone was caught up in amateur drama, largely as a director, while Walton, minister's son that he was, concentrated on religion. He had been prompt in seeking out a Methodist church to attend – he often went twice on Sundays – and in time he also became an officer of the Cambridge Wesleyan Society, but his religious interests were broad. If a well-known preacher of whatever denomination happened to be visiting the town Walton would usually make the effort to attend. Among those he heard was Sir Oliver Lodge, one of the grand old men of British physics but also a leading supporter of spiritualism. On another occasion he attended a Muslim event with a colleague from India and in a letter afterwards showed a scientist's curiosity about Islamic practice: 'During the lunar month of February he fasts from sunrise to sunset, but he does not know what he would do if he were inside the Arctic Circle during that month.'[10] Such interests were much more the norm than we might imagine today; in the 1920s and 1930s not only was church attendance in Cambridge far higher but religious clubs and other gatherings were very popular, frequently more so than political and even musical societies.

As for sport, Walton was a cross-country runner and spent most winter Saturday afternoons pounding the lanes around Cambridge with the university Hare and Hounds club. In summer he preferred to swim and he was to find an occasional swimming companion – perhaps an unlikely one – in the theoretician Paul Dirac. Still in his twenties, Dirac was Fowler's most successful protégé, having come from Bristol University with degrees in maths and electrical engineering and then, on arrival in Cambridge, swiftly made himself a master of quantum mechanics. He had already, in 1929, done the work that was to win him the Nobel prize in 1933, just as he had established a habit of extending abstract thought to everyday life that would entertain colleagues for decades. A popular example was the occasion when he found himself in company with a woman who was knitting: after watching in silence for a while he pointed out helpfully, 'You could also do it in a different way,' and began to explain. 'Of course you can,' the woman interrupted indignantly, 'that is purl.'[11] She had no idea that Dirac had deduced this purely by logic, and in a matter of minutes. What the other-worldly theoretician and the down-to-earth Walton discussed on their swimming outings we do not know; perhaps it was electrical engineering. Walton also occasionally went punting with colleagues on Saturdays, and sometimes would play tennis at the Cockcrofts', while on Sunday afternoons he regularly took a long walk, with a companion if he could recruit one (Allibone was sometimes willing), or failing that on his own. On returning home in the evening he would write letters. One thing he never did on Sunday was work.

Cockcroft's life was strikingly different, not least because he had a wife: Elizabeth Crabtree, the Todmorden girl he had fallen in love with in his teens and to whom he had written so faithfully from the trenches. Theirs had been a long courtship, for even after his safe return from war he felt unable to propose before achieving a suitable measure of financial security. 'Cambridge has a way of running off with money,' he explained to her. 'I don't think you would enjoy life in mediocre rooms with an insufficient income, in a place like this.'[12] Only in 1925, when he was established at the Cavendish with steady money coming in from a college scholarship and some consultancy work he did for Metro-Vick, did the wedding take place. It had been nine years coming, he reminded her in his last letter to her as a bachelor, but there was only good in that: 'I'm sure we've got to appreciate each other more deeply as the years have rolled on so that Wednesday's ceremony will only seal what has been developing so long – a real companionship and a bond of love which must endure.'[13] And so it proved. The most undemonstrative of men in the laboratory, he was at home a loving husband who, whenever they were apart even for a couple of days, would always write a few lines to the woman he addressed simply as 'Beloved'. Both were keen to have children and early in 1927 a son was born who was named, like his father and grandfather, John. Though they toyed with a second name of Rutherford they decided against it and eventually the baby came to be known as Timothy.

By early 1929 the Cockcrofts had prospered sufficiently to move from their small house by the Cam to a newly built, architect-designed property in Sedley Taylor Road, on the southern outskirts of town. Detached, distinctive and rectangular, with modern if not avant-garde features, the house had a large garden complete with its own tennis court – conspicuously grander accommodation than was enjoyed by the average Cavendish researcher, even after five years' work. Cockcroft also acquired his own car, a Morris which he picked up cheap and which became known to the family as Jumbo and to everyone at the laboratory as the Five Pound Car. (Challenged once, after a colleague bought a similar model for £3.10s, he quipped: 'I paid extra for mine to go straight.')[14] When he wanted, he could now race up to London in a couple of hours with two or three colleagues, 'through fresh, green country lanes',[15] park the Five Pound Car somewhere in St James's and attend an event at the Royal Society or the Royal Institution. Such comforts owed little or nothing to the support of parents and everything to hard work, for jobs and responsibilities seemed to stick to Cockcroft like filings to a magnet. Besides his own Ph.D. research and the new accelerator project, he was still carrying out studies for Metropolitan-Vickers and expending many, many hours on the development of Kapitza's laboratory. From the end of 1928 he was also a Fellow of his college, where he was recruited to organize the rewiring, and at the beginning of 1930 he was appointed a 'demonstrator' by the university, which meant lecturing to undergraduates. And that was probably not all: among his papers, for example, is a log book of the Cavendish library, suggesting that for a time at least he also took that under his wing. As one who knew him in this period remarked: 'The general impression was that this man was performing *very satisfactorily* about two and a half full-time jobs.'[16] Though money was coming in the sums were by no means princely and it took careful financial planning to make the most of it. This was another hallmark of the man: each year he and Elizabeth drew up a budget for the twelve months to come, and any surpluses were carefully invested.

They were popular and entertained frequently. The Oliphants were their closest friends, another young married couple of about the same age. 'Bones' Allibone was a regular visitor to their home, as were Metro-Vick staff on visits from Manchester, and every few months Walton would be invited to join a lunch, a tennis party or a dinner party at Sedley Taylor Road – no doubt a welcome change from the bachelor life. Cockcroft and Walton, however, were not destined to become close friends. On the surface they had a good deal in common but the similarities sometimes masked deeper differences of character. While Cockcroft had also been brought up a Methodist, for example, he was not a churchgoer unless it was to admire the architecture or hear the music; in fact he showed so little trace of his religious background that Walton was unaware of it for years. And their shared liking for sport found expression in very different ways, with Cockcroft preferring team games. As opening bat for the Cavendish scientists in their annual cricket fixture against the technical staff – another popular ritual in the laboratory calendar – he was frequently a matchwinner, while he was also a sturdy right back for the laboratory's mixed hockey XI.

Late in 1929 calamity befell the Cockcrofts. Young Timothy, who was not only the first child in the family but also the first grandchild on both sides, had grown into a beautiful and much adored boy. A photograph survives of him at play with a toy railway signal set, his expression intent, his blond hair neatly brushed and his cheeks and legs chubby – 'such a lovely flower',[17] as one grandmother put it. Cockcroft the scientist was entranced by the boy's developing imagination and curiosity. But 'Timmie' was prone to illness and was eventually diagnosed as having asthma, an even greater cause of worry then than it is today. His parents did all they could: at the house in Sedley Taylor Road, for example, they had an especially large window built into the drawing room to admit ultraviolet rays, then thought to be good for asthmatics. At the beginning of November, however, at the age of two years and nine months, Timmie suddenly fell seriously ill and within a few days was dead. The family's anguish is captured in a letter from Elizabeth's parents. 'The longing to touch and see and kiss our little lad is almost too much,' wrote her mother, while her father declared: 'I think no other little lad could have drawn so many messages of sympathy – to see him once was to impress his image upon you.'[18] Walton was among the Cavendish staff who attended the funeral but as another colleague remarked: 'Dr Cockcroft was not a man to whom it was easy to say any words of sympathy.'[19] For years they kept the flowers fresh at the boy's grave.

Cockcroft was no doubt fortunate during this period in having more than enough work at the laboratory to distract him, and by this time he occupied a curious position in the hierarchy. No longer a research student, he was not yet one of the big wheels like Chadwick or Aston nor did he rank among the younger men obviously destined for great distinction, such as Blackett or Oliphant; his scientific work did not dazzle as theirs did. And yet he was an important figure. People relied on him and he did not let them down; more, he took initiatives which improved the lives of all – it was he, for example, who recruited the lab's first secretary, Joyce Stebbings, beginning a process by which official paperwork came to be put on a professional footing.

That Cockcroft was more than just a competent administrator, however, is clearest if we try to see him through Rutherford's eyes. He had only just arrived at the Cavendish when he became, in a way, midwife to its most expensive and elaborate project, the Kapitza lab, and he had helped make that a success. Then in 1928, deploying his theoretical gifts, he had initiated the accelerator experiment, second only to the magnetic equipment in cost and scale. Rutherford trusted him with both, indeed with the second he backed Cockcroft even on what he himself considered, as we shall see, to be a long shot. This suggests on the professor's side a particular insight: that Cockcroft was a man capable of moving the Cavendish towards the next age of physics. Having reconciled himself to the idea that big machines would be important in that age, the director was relieved to have someone competent to introduce them to his laboratory.

This meant that the accelerator project, and Walton with it, also enjoyed an unusual status. As with Cockcroft, the Irishman's scientific record to date was hardly thrilling: he had been at the laboratory more than two years and had run two unsuccessful experiments and published one fairly modest paper. A Ph.D. thesis was nowhere in sight. When in mid-1929 the time came to renew his 1851 award for a third year he was in the thick of the rectifier work and still had little to show for his time in Cambridge, but Rutherford stepped in with a strong endorsement. 'I consider Mr Walton to be an original and able man who has tackled a very difficult problem with energy and skill,' he wrote. 'I am thoroughly satisfied with his work.'[20] The extension was approved. Rutherford was determined to see this work done and he made sure they were left to get on with it.

With Cockcroft so busy it was inevitable that the task of building the apparatus and carrying out the many trial-and-error experiments fell largely on Walton. This suited both men and the project benefited too: Cockcroft lacked the patience and dexterity of the natural experimenter while Walton had both in abundance – Allibone recalled later that he had such a good eye and steady hand that he could not only mend watches, but make new parts for them. He loved the work, loved its combination of discipline and imagination, loved the process of eliminating what was wrong until he found what was right and loved the tools and the materials of the workbench. Unlike many in the Cavendish he had long experience of workshop practice, accumulated throughout his boyhood at home. He did not do all of the testing and construction but he did most, and it was when problems arose or matters of strategy were to be decided that the two worked most fully as partners. Cockcroft was the senior man but Walton's mathematical ability stood him in good stead and the relationship had a healthy balance. The Irishman would say later: 'I can't recall a single instance of the slightest thing you might call a minor row.'[21]

Rutherford's direct patronage, combined with Cockcroft's relative seniority, also helped free the two scientists from the worst constraints of the Cavendish budget. Every memoir of this period includes some account of the desperate scarcity of equipment and also of the almost sadistic meanness of Lincoln, who kept the key to the stores. If a researcher asked for six brass screws he would be given four steel ones; if he sought mahogany he would be given some lesser timber lopped off a piece of broken furniture; if he wanted copper wire it would be carefully measured to half the length he requested. Legend has it that one student who applied for a short piece of metal tubing was directed by Lincoln to the courtyard outside and told to saw something off a bicycle. A man of Edwardian bearing, with a waxed moustache curled up in points on each side, Lincoln was a Cavendish fixture who had joined the workshops as a boy so small he had to stand on a box when he operated the lathes. To a degree he may have been enforcing the disciplines of that earlier age, when equipment tended to be more modest, but he was also trying to hold down costs and in this he had the support of Rutherford, who liked to tell students who complained about scarce resources that he himself could do research at the North Pole if required.

This was one of the professor's areas of inconsistency and also one of the few points on which his intimates were later prepared to criticize him. Such was his standing in the university, the academic world and the country that, by general consent, he could have raised far larger sums than he did to support the work of the laboratory. Chadwick even told the story of a generous offer of funding made to Rutherford indirectly by some leading industrialists, an offer which he simply failed to follow up. On occasion, as with the funding for the Kapitza lab, which came from the government, or the £500 university grant for Cockcroft and Walton's transformer, he was prepared to apply for fairly large sums, but as a rule he preferred not to. Chadwick believed this reflected not so much a distaste for begging as an anxiety about creating expectation, and he cited a conversation they once had on the staircase down from the Tower room, where the laboratory's radium was kept. The Cavendish had only a small quantity and Chadwick remarked to the professor: 'What a pity it is that somebody didn't give you a gram of radium, as the women of America had given to Madame Curie.' Rutherford replied: 'My boy, I am damned glad that nobody ever did. Just you think: every year I should have to think of the import of what I had done with a whole gram of radium, and I should find it impossible to justify it.'[22] In this he was quite mistaken – if anybody, after Marie Curie, was capable of putting radium to good use it was he – but he feared the burden of expectation. Far better, he believed, to produce surprises with small resources than live in the struggle to justify large ones. Besides, there was also the North Pole factor: he had made his name with simple apparatus and he still believed simple apparatus was best. While it was natural for young researchers to want more it did them good to make do.

Cockcroft and Walton suffered relatively little from this parsimony even though their appetite for hardware must have been considerable. Lincoln was toughest on the younger scientists and no doubt would readily have shown Walton his grudging side, but he does not seem to have been given the chance. It was Cockcroft who took on the task of scrounging for materials and by Walton's later account he was extremely good at it. Not only could be draw on Cavendish stores but he also knew his way around other laboratories and sources of equipment in the university and elsewhere. One example of his resourcefulness was the Apiezon pumps but another comes from a few years later in his career, when he took on responsibility for the upkeep of his college buildings. Friends reported that he continually had his eye out for supplies of old bricks that would match the ancient fabric of St John's and sometimes on a country drive he would halt to inspect a pile of rubble on the roadside and then, after a little haggling, buy it from the farmer on the spot. The same efficiency and the same sharp eye brought great benefits to the accelerator project.

The two men passed a milestone in December, with the delivery of the transformer from Manchester. Having safely negotiated the corridors and the doorway, this large black object, resembling a tall pot-bellied stove, took its place against the brick wall of the workroom. The only hitch was that it turned out to be heavier than expected and so the floor had to be reinforced. Over the weeks that followed, the rectifiers were also assembled for the first time in the chosen arrangement, the two metre-long glass tubes held horizontally at chest height on rods of the insulating material Bakelite. Where they met there was a drum-shaped centre joint and downwards from this ran a third bulb, leading to the vacuum pump. On either side of the rectifiers, like the mouths of two great silvery trombones, were fixed the large corona shields. Though big by the standards of Cavendish apparatus and certainly dangerous when live, with its gleaming metal parts and highly polished glass bulbs it all had a certain beauty. Now it was ready to be tested.

9. Other Ideas

Cockcroft and Walton were not the only physicists working on a fresh approach to the nucleus. Even in their own laboratory there were groups developing alternative ideas while in the United States and Germany several other projects were taking shape, some of which would provide direct competition for the Cambridge pair. Of all the possibilities under investigation perhaps the most straightforward was the attempt to develop automatic particle counters. A machine that could detect and log particles, it was clear, would not only take most of the drudgery out of nuclear bombardment – all that waiting in the dark, all the eye strain – but it would also remove the subjective element from the experiments, so that the risk of further disputes would be greatly reduced. Better still, it could extend the potential of the technique. If the machine was more sensitive than the human eye and capable of operating for long periods then new kinds of experiment would become viable without having to replace the traditional particle sources. The whole business of disintegration, in other words, could be moved forward a step or two. Several scientists had turned their minds to this challenge and one of them worked in the very next room to Cockcroft and Walton.

Eryl Wynn-Williams had come to the Cavendish in 1925 from the University College of North Wales in Bangor and was set to work on a study of short electrical waves. A year later, however, he read a paper by a Swiss scientist, Heinrich Greinacher, which gave him another idea. Greinacher suggested that it should be possible to detect particles such as protons and alpha particles with the help of new electrical tools known as 'thermionic valves', which had been developed by the radio industry to pick up faint signals in the airwaves. As one of the Cavendish's amateur radio buffs Wynn-Williams was familiar with this technology and immediately saw the merit of the idea, so in his spare time he began building his own version of the apparatus. The trick was to exploit the tendency of fast-moving particles to cause a minor electrical disturbance when passing through a gas. As they travelled they bumped into some of the gas molecules and atoms, knocking out the occasional electron and so leaving a trail of 'ions', or broken atoms which have lost their electrical balance. Although such disturbances were feeble and fleeting in character, Greinacher said that several thermionic valves, suitably arranged, would be able to amplify the electrical effect to the point where it could be made to illuminate a light bulb or cause a click in a telephone receiver. Wynn-Williams had no trouble assembling such a device and sure enough it worked, clicking away merrily in response to a stream of alpha particles from radium. Rutherford was impressed and used the new toy with great success at public lectures. Unlike zinc sulphide scintillations, which were visible only under a microscope, the click–click–click of Wynn-Williams's new counter was something an entire hall of people could appreciate at once. It was a pleasing new stunt.

The next step was to transform what was a primitive apparatus into an instrument that could be used in research, and this task Rutherford assigned jointly to Wynn-Williams and F. A. B. Ward, a young Londoner from the same 1927 intake as Walton. They started by providing a better way of recording the counts. Clicks in a loudspeaker, appealing as they were in public display, were not intrinsically an improvement on scintillations so far as the experimenter was concerned; what was needed was an objective record. By the spring of 1928, therefore, the two scientists had replaced the clicks with a photographic system in which each amplified impulse, instead of making a sound, caused a mirror to vibrate very slightly. A fine beam of light was played on to the mirror and reflected on to a slowly moving strip of photographic paper, giving a continuous line. Each tiny vibration of the mirror caused a jump in the line, which the two men called a 'kick'. Each kick recorded on the paper, therefore, marked the passing of a single ionizing particle through the gas.

Considering that the objective was to create an efficient and convenient device for recording particles, Rutherford's faith in the work at this stage was impressive, for the apparatus was anything but convenient. The chief problem was its vulnerability to external disturbance. Even when much of it had been suspended from the ceiling on long springs and other parts were cushioned on rubber sponges or shrouded in protective screens, the slightest thing could wreck a counting session. Noise and physical vibration, rarely absent in Rutherford's Cavendish, were serious hazards and there could also be trouble if anyone in a nearby workroom was tinkering with quite modest voltages (truces had to be agreed with the high-voltage men). Scarcely less troublesome was the photographic recording process, which in a standard experimental session of thirty-five minutes would generate a strip of paper more than 100 metres in length. This had to be wound carefully on a rig and then developed in sections – a messy business – and after that the job still remained of counting the kicks by eye. We can picture Wynn-Williams and Ward tiptoeing around their Heath Robinson construction in constant fear that the turning of the big wheel for the film would disturb the little mirror, or that a door would bang somewhere or that a light switch might be turned on in the corridor outside. Though it produced good-quality results when conditions were right, operating it was such a tightrope walk that most researchers would surely still have preferred the tedium of scintillation counting.

One summer's day in 1928, while this development work pressed ahead, a brown-paper parcel with a German postmark arrived on Rutherford's desk. It had been sent by his old friend and former colleague Hans Geiger and contained a prototype of the device that would make Geiger's a household name: his own, rather different particle counter. Rutherford had every reason to recognize it because it was a distant descendant of a temperamental apparatus he and Geiger had built together more than twenty years earlier in Manchester. As with Greinacher's technique it exploited the ionization effect of fast particles in gas, but it did not use valves. Instead Geiger amplified the effect directly inside the little chamber containing the gas. He inserted live electrical terminals so that when a passing fast particle knocked electrons out of an atom, the resulting positive ion was attracted towards a negative terminal. This attraction was made so powerful that the ion rushed towards the terminal and as it did so it struck other gas atoms, breaking them up and creating more ions, which were also drawn towards the terminal. This process multiplied itself until it became an avalanche of ions so intense it could be translated directly into a click, a flash or, as in the Wynn-Williams arrangement, a 'kick' on a photographic line.

Which technique was better, the valve counter or the Geiger? The German, who had been working on his device for much longer, was ahead in most respects: his counter was compact and practical to use as well as extremely sensitive. Not only did it detect alpha particles and protons but it also picked up electrons and even gamma rays – an area where the scintillation screen had been very weak. Little wonder that the Geiger counter was soon adopted with delight by many researchers, and especially those studying electrons and gamma rays. But it had its shortcomings. One was that if several particles arrived in quick succession it would pick up the first but miss the others entirely. This was because it took a moment to recover from each avalanche. Another was that it could not distinguish between particles of different kinds: every incident in the tube, whether small or large, ended in an avalanche of exactly the same scale and so everything from alpha particles down to gamma rays registered simply as one click, one kick or one flash. The Wynn–Williams device, though it was at a much earlier stage of development and was less effective when it came to smaller particles, promised to be stronger in those other areas. Its shorter recovery time gave it the potential to count individual particles even in the midst of a rapid burst, while it was also able to register different events in distinctive ways – an alpha particle kick on the photographic paper, for example, was visibly taller than a proton kick. This second capability would be extremely valuable in a great deal of nuclear research work. In short, while the Geiger counter rightly won many admirers across the field of atomic studies the valve device appeared to have the greater long-term potential for physicists whose primary interest was the bombardment of the nucleus.

In another year, by which time Rutherford had presented his friend Geiger with a Royal Society medal in recognition of the success of his counter, Wynn–Williams and Ward had made further quiet progress. They rebuilt the electrical sections in more sturdy form, isolating the most sensitive parts so that they could dispense with the cradle of springs, while as an alternative to the photographic process they introduced a mechanical counter like a taxi meter. Thus in December 1929, a week or so after Geiger was honoured at the Royal Society and around the time when Cockcroft and Walton were taking delivery of their transformer, Wynn-Williams and Ward were beginning to test their latest machine in a full-blown experiment run by Rutherford himself, helping him to measure alpha-particle ranges. So promising were the developments in automatic counting that a feeling was spreading through the Cavendish that optical scintillation counting would soon be a thing of the past. Bern Sargent, a young Canadian in the Nursery course, was so convinced of this that he wanted to avoid the old technique altogether, and when he and another student were set a scintillation exercise the two even considered faking their results to ensure they would not be asked again. 'When our efficiencies worked out at only eighty-five per cent each,' Sargent admitted later, 'we decided that cheating was not necessary.'[1] Such confidence in the automatic techniques was a little premature, partly for the humdrum reason that new methods were always slow to replace old ones in the Cavendish: Geiger counters arrived in only ones and twos over the years that followed, as resources permitted, while Wynn-Williams and Ward needed more time to create a finished product. None the less it was true that the days of scintillation counting, that happy product of Crookes's serendipity which had become such a burden on the researcher's back, were now numbered.

Moving in tandem with these advances, a small number of scientists were having fresh ideas about the radioactive materials used in nuclear research. They were looking for a way to get more out of the natural particle sources without having to make the radical switch to artificial acceleration that Cockcroft and Walton were proposing. More intense natural beams, when combined with the automatic counters, might just be enough to bring the fly in the cathedral into reach, they argued. Once again some of this thinking was going on in the Cavendish, and this time it was James Chadwick who was doing it.

Sharp-featured, lean and with an ascetic air, Chadwick carried all his life the scars of early hardship. Born in a country village near Manchester in 1891, he was the son of a cotton spinner who moved to the city in pursuit of something better and ended up if anything worse off. 'We were very poor,' Chadwick said later. 'Well, poor anyhow.'[2] Young James, happily, was very bright and by dint of hard work and scholarships eventually earned himself a place at Manchester University. He walked the four miles there every day, studied or played chess while the better-off students had lunch and in the evening walked home again; broke and shy, he allowed the cultural and social life of the university to pass him by. Like so many others, however, he was inspired by Rutherford to pursue a career in research and in 1913 he landed the apparently golden opportunity to study in Berlin, a city then at the height of its powers in physics. Working there under Geiger he encountered some of the great personalities in his field, most notably Einstein, who stopped to chat with him once during a visit to the laboratory and produced one of his famous epigrams. A question from the Englishman about some odd results in his experiments elicited the response: 'I can explain either of these things, but I can't explain them both at the same time.'[3]

The outbreak of the First World War trapped Chadwick in Germany and the four years he spent as an internee plunged him back into conditions at least as desperate as he had known in childhood. 'Physically, it nearly killed me,' he recalled.[4] He managed to conduct some experiments (and also taught the principles of radioactivity to a fellow prisoner, Charles Ellis, who would go on to become a pillar of the Cavendish), but his health and nerves were so cruelly affected that for the rest of his life he believed himself to be almost permanently at the brink of collapse. On returning to Britain in 1919, once again penniless, he rejoined Rutherford and followed him to Cambridge as a sort of scientific dogsbody. Better things followed, however, for before long he was in the thick of the work on artificial disintegration, first as Rutherford's junior partner and then in his own right, while his administrative efforts were recognized with a promotion to assistant director. Now at last he had stability, professional success and even modest prosperity, and in 1925 he married the daughter of a wealthy Liverpool banking family.

Chadwick knew Rutherford better than any other scientist and in his later years, after the great man died, he would become one of the guardians of his reputation, editing the collected works and lecturing on his life. Yet even while Chadwick expounded on Rutherford's genius he could not help betraying the stresses he himself felt as those Cambridge years passed. There were many points of friction. Chadwick wanted to reform the laboratory so that, instead of students always having to build apparatus for every new experiment, a few of the more complicated sets would be maintained permanently for whoever needed them. Rutherford said no. He wanted to experiment with high voltages as early as 1922, but again Rutherford said no. He struggled to stock the laboratory with tools such as vacuum pumps, but while Rutherford could have raised more money he would not try. Always such memories were clothed in respect, but the frustration is evident.

By the late 1920s Chadwick was approaching his forties and increasingly restless. Where once he had been in awe of his tempestuous boss, now when they disagreed he was less inclined to bite his tongue. One example was his response to Rutherford's 1927 paper on the structure of radioactive atoms — the 'tug-boat' paper that Gamow would see in Göttingen. Chadwick recalled later: 'I was working with him [Rutherford] at the time; he talked to me about it. But I had not much opinion of it. It seemed to me to have very little, to put it mildly, significance. He was rather annoyed with me at the time.'[5] As for Rutherford, though we have no account from him of the relationship there are grounds to infer that he too felt the change of mood. Because he was conducting less research his dealings with Chadwick came to be concerned more with administrative matters, and with Chadwick now an established figure known beyond the walls of the Cavendish it is easy to see how their occasional disagreements would take on more significance for the older man; willy-nilly he had to accept a greater equality between them. It would be wrong to exaggerate this; there was no feud, but after ten years of collaboration the two were developing different priorities.

When in 1927–9 Rutherford grew keen on seeking artificial means of accelerating particles Chadwick, despite his earlier advocacy, did not become involved. In general terms he supported the initiative but there is little or no trace in the record, or in the recollections of Cockcroft, Walton, Allibone or Rutherford, of him taking a direct part in the management. Instead Chadwick found a new direction of his own and carried a number of young researchers along with him. The ultimate aim remained the same as Rutherford's — the penetration of the nucleus — but the approach was very different and more in tune with developments in continental Europe. It involved deploying automatic counters in the search for nuclear effects, but it also required a second new ingredient, an unfamiliar one. The idea seems to have broken the surface at a conference on atomic physics that Chadwick organized in Cambridge in the summer of 1928.

This was a leisurely affair, in keeping with the season, and student gossip suggested that the true motive behind it was to check that nothing important had been left out of a textbook that Rutherford, Chadwick and Ellis were then completing. The main European schools of experimental atomic physics were represented: Hans Geiger came from Kiel, where he was now based, and Walther Bothe from the Reichsanstalt in Berlin, while Lise Meitner represented Berlin's other leading laboratory, the Kaiser Wilhelm Institute (she worked there with yet another former collaborator of Rutherford's, Otto Hahn). Geiger, Bothe and Meitner were all of an age with Chadwick and Ellis: about forty. From Paris came two younger scientists, the new generation at the Radium Institute founded by Marie Curie, who was in her sixties and increasingly frail. Frédéric Joliot, not yet thirty, was Madame Curie's personal assistant and beginning to make a name as an experimenter, while Irène Curie, a couple of years older, was the founder's daughter and had published a string of important papers, mostly on alpha particles. These two, Joliot and the younger Curie, were married to each other. Some of the other British universities were represented, but notably absent from the gathering was the Vienna school. Whether Pettersson and Kirsch were not invited or whether they chose to stay away is not known.

It was in the course of the mostly informal conversations on those summer days that Chadwick, Meitner and Bothe fell to discussing the potential of polonium as an alternative to radium for use in disintegration experiments. Like radium, polonium had been discovered by the Curies in 1898 (it was named in honour of Marie's homeland, Poland, then under foreign rule), but while radium was extremely scarce polonium was 500 times scarcer: 10,000 tonnes of ore were needed to produce a single gram. It is easy to see why radium had always been preferred as a research tool. But polonium had another important characteristic: in the course of its radioactive decay it emitted no electrons or gamma rays, only alpha particles, and while these were not quite so swift as radium alpha particles they were vastly more plentiful, in fact 5,000 times more so. To anyone interested in the results of nuclear bombardments at the end of the 1920s this represented a substantial advantage: polonium had the potential to produce far more data, far more quickly. Instead of polling rare passers-by on a Yorkshire moor, working with this more prolific element would be like sampling pedestrians in a busy shopping street. And with the arrival of automatic counting it was becoming possible for the scientists to cope with flows of particles that would have simply swamped the old scintillation screens. The way was opening up. Combining the more plentiful ammunition from polonium with the more sensitive observation that could be achieved with automatic counters offered the prospect of a closer scrutiny of the nucleus than had hitherto been possible.

The problem for Chadwick was that while the Paris institute and the two Berlin laboratories had stocks of polonium, the Cavendish did not. When he mentioned this fact Lise Meitner generously offered to send him a sample and when it arrived that autumn he set a research student to work with it immediately. Frustratingly there was not quite enough to get meaningful results but Chadwick's appetite was whetted and he began to look out for additional supplies. While he searched, however, Walther Bothe was already hard at work at the Berlin Reichsanstalt. An adept user of the Geiger counter, Bothe had begun a series of experiments with polonium that would soon cause great excitement in the small world of nuclear physics.

★

There was, therefore, more than one tempting new path towards the nucleus as the 1920s drew to a close, albeit that results still seemed to be some way off. And even on the path of high voltages, now preferred by Rutherford and trodden at steady pace by Cockcroft and Walton, other scientists were beginning to appear. One of them, indeed, had been following it for some time. As early as February 1925, when Walton was still in Dublin and Cockcroft a Cavendish newcomer, a young man on the other side of the Atlantic was already thinking about accelerating particles for use in nuclear research. Merle Tuve was a Norwegian-American from the small South Dakota town of Canton who worked his way across to the emerging physics research laboratories of America's east coast. Things electrical had excited him since childhood, when he and a boy from across the street had built a radio set that picked up signals from as far away as New Jersey. Tuve studied electrical engineering at the University of Minnesota, did a little research there and then headed first to Princeton and then to Johns Hopkins in Baltimore, Maryland. His main field was the study of the electrical characteristics of the upper atmosphere, using radio waves, but somewhere at the back of his mind the idea of particle acceleration had taken hold. In 1923 he first discussed it with a theoretician friend, Gregory Breit, and it took firmer shape after he heard Rutherford lecture during an American tour the following year. By 1925 he had decided that he wanted this to be his big project and like so many others he felt that the best place to do it must be the Cavendish Laboratory, so he got hold of some forms to apply for a grant to go to England for a year. Just as he completed the application, however, he received an alternative offer.

His upper-atmosphere studies were a collaboration with Breit, who did not work in Johns Hopkins but at the Carnegie

Institution in Washington, DC, some forty miles away, and Breit had persuaded the institution to offer Tuve a research post. The first response was dismissive. 'I want to go and do high voltages with Rutherford,' said Tuve. 'Somebody has got to make high-speed protons, artificial alpha particles. We've got to speed up these things and there's no reason why one can't do it.'[6] Breit and his colleagues were not so easily put off. The upper-atmosphere study was important and advancing well, they argued, so it would be a shame to bring it to a halt, and as for high-speed protons he could hardly hope to achieve that in one year in England, so why not come to Washington and take his time? Since Tuve had no idea yet whether the Cavendish would accept him he took the bird in the hand and in 1926 transferred to the Carnegie's sonorously titled Department of Terrestrial Magnetism.

Rutherford might have welcomed Tuve – at the age of twenty-three the American had done good work and could command impressive references – but whether the Cavendish director was ready at that time to support an attempt to accelerate protons is doubtful. As it turned out, the Carnegie Institution was not terribly keen either: the department had a much stronger tradition of fieldwork than laboratory research and its resources in money and technical support were limited, so high-voltage research had to take second place to the less expensive upper-atmosphere studies. Tuve none the less managed to build a Tesla transformer with which he produced first 3 million volts and then, with further development, 5 million – a record at the time and sufficient to earn him an honourable mention in Rutherford's 1927 Royal Society talk. But there was still a long way to go before the American could apply such voltages to an experimental tube in a controlled fashion, and the pace was slow. It picked up dramatically after Breit paid a visit to Europe in late 1928.

Officially he was sent to study under Pauli in Zurich but he took the chance to make the typical theorist's tour, visiting Vienna, Leipzig, Berlin and the Netherlands, meeting Heisenberg, Schrödinger, Ehrenfest and Gamow's up-and-coming friend Wigner, as well as a number of leading experimentalists. Having spent the previous few years breathlessly observing the eruption of quantum mechanics from afar, the American was shocked at what he found. Progress at the very heart of theoretical physics appeared to have slowed dramatically, advances were being made only in secondary fields and even the best minds were held in check by a small number of basic problems. Reporting on this malaise to his superiors in Washington, Breit made a bold but to us familiar diagnosis: the cause, he wrote, was 'the scarcity of experimental information about the atomic nucleus'. And he offered a prescription: 'Very helpful data for the solution of the remaining problems in theoretical physics can be obtained by experiments on nuclear disintegration (this is brought out very strikingly by Gamow's new work on the theory of nuclear disintegrations).'[7] The experiments he had in mind were high-voltage ones and so urgent did he feel the need to be, and so promising the opportunity, that he broke off his stay in Europe early to head home and see what answers Tuve's equipment could provide. He was back by mid-January 1929 (just when Gamow was visiting Cambridge to discuss similar ideas with Rutherford and Cockcroft) and very soon the attack on the nucleus was at the top of the agenda for the Washington laboratory.

The apparatus that Tuve and his colleagues Larry Hafstad and Odd Dahl had created bore little outward resemblance to the equipment being set up at the Cavendish by Cockcroft and Walton. Most strikingly the Washington set-up operated under pressure inside a large, closed tank of oil, a measure which, though messy, greatly reduced corona problems and increased the tolerance of the tube. The tube itself was also very different in design, being made up of a number of Pyrex bulbs resembling those later used in coffee percolators, fitted together and operating horizontally. The first few months after Breit's return saw them testing a six-bulb tube and when that proved successful they moved up rapidly to a fifteen-bulb model. That summer they managed to apply 1.4 million volts to it without breakdown. Impressive as this was, however, it did not mean they were on the point of disintegrating nuclei, for their equipment was not well suited to that task. The problem, as Rutherford had diagnosed, lay with the power source, for while the Tesla coil offered a cheap and relatively easy route to high voltages its output came in rapid, drastic peaks and troughs, so that it worked at maximum voltage only for about one-thousandth of the total operating time. It could never produce the smooth, homogeneous stream of fast particles best suited to bombarding nuclei; the best that could be hoped for was muddled, fleeting surges that would be very difficult to work with. So while there was a certain amount to be learned by using the apparatus as it stood, the Washington team knew that to move to the next stage they would need a different power source, and they were on the lookout. When a Kentucky firm quoted a price of more than $2,000 for a basic 200,000-volt transformer, however, the Carnegie Institution management shook its head. For the moment Tuve was stuck with a Tesla machine he would later describe as 'our albatross'.[8]

Besides Tuve and his colleagues there were others dreaming of accelerating particles by the use of high voltages, and it was a group of young physicists at the Friedrich-Wilhelms University in Berlin who adopted the most exotic approach. There were three of them, Arno Brasch, Fritz Lange and Kurt Urban, and in 1927 they set out for the high Alps with the aim of capturing natural lightning. This was no hare-brained scheme borrowed from the *Frankenstein* of the cinema; they were capable, sensible researchers supervised by the highly reputed Walther Nernst and supported by the engineering firm Brown Boveri, so the science was well grounded. And lightning experiments, dangerous as they were, had a pedigree that went back to Benjamin Franklin. The initial aim was to investigate whether these huge natural electrical surges – lightning flashes are typically of the order of 10–20 million volts – could be tapped in a controlled way, with the possibility that they might be harnessed for atomic experiments as a long-term hope.

Early summer of 1927 thus found the three Germans rigging cables between two ridges just below the 1,700-metre summit of Monte Generoso in southern Switzerland. Known for its thunderstorms, the site was served by a cog railway built for tourists and there was a convenient hotel at 1,600 metres where the scientists set up their headquarters. Their first scheme involved suspending a wire net over the middle of their 600-metre-wide valley, with heavy-duty insulators on either end of the cable. The results were disappointing but the three returned the following year with an improved design involving a double cable and two metal spheres, one hung eighteen metres above the other. Sure enough, in thunderstorms they were able to observe huge sparks leaping between the spheres and they calculated that this was 18-million-volt lightning. It was an impressive first step, potentially offering them a source of power vastly greater than was available to any other scientists, but their progress was abruptly halted on 20 August 1928, when Kurt Urban was killed in a fall. Lange and Brasch decided the dangers of high-altitude work were too great and returned to the laboratory.

They did not, however, abandon high voltages, instead

applying to Allgemeine Elektrizitäts-Gesellschaft (AEG) to make use of a new 2.4-million-volt generator the engineering firm had just built for testing purposes at its Berlin laboratories. This was a 'surge' generator which produced pulses lasting just a few thousandths of a second – not unlike lightning – so the difficulties in using it for research work would be considerable. AEG were keen to assist but the commercial imperative played a role: in effect, Brasch and Lange were allowed to use the machine on condition they helped the firm develop high-powered X-ray apparatus of a kind then in demand in hospitals. This seemed to the company to offer the best prospect of a financial return. The deal was done and the two physicists plunged into a remarkable programme of experiments, creating an impressive tube of porcelain which they sank in oil and pushed to steadily higher voltages. This work earned them the admiration of Tuve and enabled AEG to put some advanced X-ray apparatus on the market, but the scientists' hopes of accelerating particles had to remain on hold.

X-rays also dominated the work of another laboratory where high voltages were available and where ground-breaking work was done on the development of better tubes. This was the California Institute of Technology in Pasadena, where a Danish-born American, Charles Lauritsen, built a machine that would greatly influence the work of the Cavendish. Lauritsen trained as an architect in Denmark and then worked for ten years as a radio technician in the United States before deciding in 1926 that he 'would like to go to school again'.[9] He drove to Pasadena, enrolled as a graduate student and within a short time was working on the problems of applying high voltages to glass tubes. Such voltages were readily available at Caltech because of a deal with Southern California Edison, the state's biggest power company, under which the institute got first-rate apparatus in return for help in designing transmission equipment. Lauritsen thus had up to 2 million volts of alternating current more or less on tap. For his tube he adopted an approach quite different from those chosen by Tuve, by Cockcroft and Walton and by the Germans, employing straight glass cylinders of a kind then familiar to most Americans. At petrol stations in those days the pumps included glass sections which allowed customers to see the petrol passing through on its way to their tanks – a means of showing them they were really getting something for their dollars. The sections were about thirty centimetres wide and seventy-five in length and made of very thick glass, far sturdier than the rugby-ball bulbs that Cockcroft and Walton chose. Lauritsen, working with a younger man, Ralph Bennett, placed three of these one on top of the other, secured them inside a timber gantry like an old-fashioned oil well and then fitted an elaborate structure of corona shields at each joint to prevent sparking. As early as December 1928 the Caltech pair were able to publish a paper in the leading American journal, *Physical Review*, announcing that this apparatus functioned satisfactorily at up to 750,000 volts.

Lauritsen had solved many of the problems that Cockcroft and Walton faced even before they began. Working in the well-endowed, well-equipped Pasadena laboratory he was able to apply more than twice the voltage the Cambridge men were seeking (albeit an unrectified, a.c. voltage) to an experimental tube that was as ingenious as it was physically impressive. Yet there was no plan to use this set-up for nuclear research because Lauritsen and his superiors, like AEG in Germany, were primarily interested in high-energy X-rays. These were being increasingly used not only for the familiar diagnostic purposes but also in radiotherapy, and in fact Lauritsen's machine soon became a medical instrument in its

The Cavendish
Laboratory, Free
School Lane entrance.
Founded in 1874, by
the 1920s the
laboratory was already
a place of illustrious
tradition in physics.

Ernest Walton, Ernest Rutherford and John Cockcroft
posing for news photographers outside the lab on the
morning of 2 May 1932.

James Chadwick, assistant director of the Cavendish. Diffident but brilliant, Chadwick mounted his own attack on the atomic nucleus.

Cockcroft with George Gamow. These two provided the essential spark for the experiment.

The courting couple: Winifred Wilson and Ernest Walton in the countryside outside Belfast following one of their occasional rendezvous in the city. The romance began with a chance encounter on a railway station platform.

The married couple: Elizabeth and John Cockcroft at their home in Cambridge. Sweethearts since John's army days in the First World War, they married in 1925 but soon suffered tragedy.

The Cockcroft-Walton apparatus in 1930. The dark object against the wall in the background is the transformer, and the two glass rectifier bulbs are arranged horizontally in front of it with a grey metal corona shield at each end. The bulb leading down from the drum in the centre links the rectifiers to the Burch vacuum pump. On the right in the foreground is the acceleration tube, with pipe-shaped anode and cathode clearly visible inside.

Merle Tuve. The first American to tackle particle acceleration, he was handicapped for years by the lack of a good-quality power source.

Robert van de Graaff with the first prototype of his generator in 1931. Tuve seized on this apparatus as the answer to his prayers.

Stanley Livingston (left) and Ernest Lawrence standing inside the giant electromagnet for their 27-inch cyclotron.

The rudimentary control table for the 800,000-volt apparatus in the former Lecture Room D, with Cockcroft at work.

The acceleration tube, with Walton in the wooden hut at its foot. The metal 'top hat' crowning the tube contains the proton supply apparatus.

The complete Cockcroft–Walton accelerator, fruit of three and a half years' labour, pictured in the *Illustrated London News* in 1932. A is the transformer, B a condenser, C the four-stage glass rectifier tower, D the acceleration tube, E the observation hut (with black curtain drawn) and F the spark gap spheres. The two unidentified towers in the background are another condenser set (next to the spheres) and the transformer for the proton supply apparatus (far right).

The core of the apparatus with which Chadwick discovered the neutron. This unimpressive cylinder, just ten centimetres long, contained polonium and beryllium and served as the neutron 'gun' in the key experiments.

Seated from left to right: Erwin Schrödinger, Irène Joliot-Curie, Niels Bohr, Abram Joffe, Marie Curie, Paul Langevin, Owen Richardson, Ernest Rutherford, Théophile de Donder, Maurice de Broglie, Louis de Broglie, Lise Meitner, James Chadwick.

Standing from left to right: E. Henriot, Francis Perrin, Frédéric Joliot, Werner Heisenberg, Hendrik Kramers, E. Strahel, Enrico Fermi, Ernest Walton, Paul Dirac, Peter Debye, Nevill Mott, B. Cabrera, George Gamow, Walther Bothe, Patrick Blackett, M. Rosenblum, J. Errera, Edmond Bauer, Wolfgang Pauli, M. Cosyns, J. Verschaffelt, E. Herzen, John Cockcroft, Charles Ellis, Rudolf Peierls, Auguste Piccard, Ernest Lawrence, Leon Rosenfeld.

The Solvay Conference of 1933. A summit meeting for leading physicists, this was a high point in the research careers of Cockcroft and Walton. Half of those pictured are Nobel prize winners.

own right, used directly on patients in the attempt to destroy tumours. Only when it could be spared from these duties was he able to carry out his fundamental investigations of X-ray behaviour, and the rest of his time he devoted to designing an improved, second-generation model.

There was one more American who would soon become a rival to Cockcroft and Walton, and by a touching coincidence he not only came from the same South Dakota town as Merle Tuve, but he was that same boy from across the road who had helped the young Tuve build a radio receiver in his basement. His name was Ernest Orlando Lawrence and in time he would become one of the most famous and admired of all American scientists. After leaving South Dakota Lawrence and Tuve had followed different paths but the two remained close friends and it was certainly Tuve who stirred in Lawrence's mind an interest in fast particles and nuclei.

Lawrence went to Yale, made a name for himself and was lured away to the University of California at Berkeley by the promise of an associate professorship, a light teaching load and generous research facilities. He arrived in the summer of 1928 and had been there six months when, one evening in the library, a paper in a German journal gave him an idea. We have met this paper before: written by the Norwegian Rolf Wideröe, it described his experiments with two devices similar to the ones Walton worked on in his first year at the Cavendish. When Walton read it he abandoned for good his attempts to build benchtop-scale accelerators, but Lawrence's reaction was quite different. It was the drawings that the American noticed – he scarcely read the text – and what set him thinking was the linear apparatus he saw depicted. Wideröe showed that it was possible to push charged particles in stages, so that they picked up speed step by step, but the process he described had the drawback that it would take a great many pushes to reach useful speeds. The necessary line of equipment, in other words, would be impractically long. Lawrence, juggling the possibilities in his head, asked himself: what if the beam were bent in a circle, so that it passed through the same pushing equipment again and again? And with that he realized something. If it were possible to achieve this, to hold the particles in a circle and still apply repeated pushes, then so far as he could see there was no limit to the speeds that could be achieved, virtually up to the speed of light. Though superficially similar to the circular schemes attempted by Walton and Widerøe, Lawrence's device would rely on different principles – and he was convinced from that first night that it would work.

For a time he was very excited, discussing it in animated terms with colleagues who, if they did not quite share his certainty, offered every encouragement. For reasons that are obscure, however, he did nothing practical about it for the best part of a year and in the end it took an illegal drink and some firm words to push him into action. Lawrence was visiting Harvard in the Christmas holidays of 1929 when he met another scientific tourist there, a professor of physics from Hamburg called Otto Stern. These were the days of Prohibition but at Stern's request Lawrence took him to a Boston speakeasy for a drink and after a few glasses of bootleg wine the American found himself describing his pet idea. Stern listened patiently, examined Lawrence's sketches on the tablecloth and eventually pronounced: 'Ernest, don't just talk any more. You must get back to California and get to work on that.'[10] Lawrence did as he was told. On his return to Berkeley he co-opted a research student to help and embarked on the construction of his new device.

By early 1930, therefore, at least five different research laboratories were working on equipment which would harness high voltages to experimental tubes, or accelerate particles to high speeds, or both. In Cambridge Cockcroft and Walton had their transformer and their bulbous glass rectifiers and experimental tube and, working from Gamow's theory, were hoping to disintegrate nuclei with just 300,000 volts. In Washington Merle Tuve was struggling with his troublesome Tesla machine in its tank of oil, aiming for millions of volts. In Pasadena Charles Lauritsen had 750,000 volts of a.c. to play with and a magnificent vertical tube, but he was pursuing mainly medical X-ray work. And in Berlin Arno Brasch and Fritz Lange, having started by playing with lightning, found themselves diverted into the same X-ray field by their need to return to the laboratory. That left Berkeley's Ernest Lawrence and his circular machine for accelerating particles. This project, once begun, would advance at a breakneck pace; the device would prove to be as potent a tool as its inventor imagined and it would soon pose the greatest threat of all to the Cavendish's hopes of being the first to smash atomic nuclei by artificial means. Lawrence's brainchild acquired a jokey nickname among his colleagues but that name would stick: it was called the 'cyclotron'.

10. Turning Point

In the early 1930s the editor of a Cambridge study guide invited Patrick Blackett to contribute an article about experimental physics. The tall, dark and handsome Blackett was among the brightest young talents of the Cavendish and the account he gave of his discipline was characteristically shrewd and original. He set the scene with some words often heard at the 4 p.m. Cavendish tea, when a researcher would amble in with shaking head, fill his cup and announce: 'I have spent two days in getting the leak in my apparatus so small that it will now take me a week to find it.'[1] Blackett's point was not that maintaining a good vacuum was often a frustrating business, although frustrating it certainly was. Joints, seals and leaks were eternal preoccupations in the Cavendish, since almost everyone was struggling with vacua. Glass connections were lovingly ground to make the snuggest fit and when two parts were joined the meeting point was sealed with wax (the favoured brand being the red wax made for the Bank of England, although being the most expensive it was difficult to extract from Lincoln's stores). For a really good seal it was best to heat the glass pieces as well as the wax, a procedure that could require a large oven, and then allow all the parts to cool slowly together, but even after this it was likely that some invisible fault would remain or would appear when the first experiment was attempted. So hunting for leaks was a job everybody knew, and most Cavendish researchers were also familiar with the grim moment when they had to give up, pull the apparatus apart and start again. In drawing attention to this Blackett wanted to impress upon young readers not only that physics involved perspiration as well as inspiration but also that to a good scientist perspiration was part of the joy of the thing. It was their calling to do what was difficult and take pains over it.

No doubt when they read this his colleagues allowed themselves a smile, for not even Blackett was capable of greeting every setback with equanimity, and yet most will have seen the essential truth. Their apparatus tended to be personal to them, not just a tool but also something to be nursed and cajoled over months and even years, and to be admired and shown off when it did its job. The successful experimental physicist, Blackett said, was a rare creature who needed skills of the hand as well as the mind, and the right temperament to boot. The balance of these qualities varied – 'the greater the experimenter's gift of understanding intuitively how an apparatus works, the less does he need actual manual skill' – and such understanding could usually be cultivated, he insisted, only through long practice.[2]

Down in their basement room in the early months of 1930, Cockcroft and Walton were able to savour their full share of these perverse pleasures, for progress continued to be slow at best. Once the transformer was hooked up to the rectifier tubes it was soon evident that the preparatory testing they had done using Allibone's Tesla tube had only limited value. The anode and cathode, the electrical terminals inside the rectifiers, were still not right and the corona shields on the outside did not do their job, which meant that the voltage was not contained and under control. With infinite care, therefore, the two men worked their way once again through the possible range of materials, shapes and sizes for all these parts. Every change to the anode or cathode involved breaking the wax seals on the joints of the apparatus, which in turn meant re-cleaning, re-heating and re-waxing before the next test. And once the vacuum pump was activated there was the familiar rigmarole of hunting the leaks, sealing them and painting over the finished joints with shellac. When everything was ready and they tried the voltage again, as often as not a spark would burst through; on a good day it might make a tiny puncture that (once located) could be repaired, but on a bad one it could shatter a whole tube. Months of this were required but step by step they managed to get things right. The best material for the anodes, they found, was nickel-plated copper, still shaped like an open-ended tin can but with a rounded lip at the open end. They settled on a cathode filament just 0.25 mm thick, while the gap between cathode and anode was made to coincide precisely with the widest part of the bulb. No sharp corners or edges were permitted on the metal parts, inside or out, since these could function like lightning conductors, and elaborate rituals were required to ensure every part was polished and cleaned to the highest laboratory standards.

These same months also saw them experimenting with another vital part of their accelerator: the proton supply. The transformer would give them their high voltage and the rectifiers would convert it to the required direct current or d.c.; that much was more or less in hand. Ultimately these would provide the powerful force of electrical repulsion that they wanted. The proton supply, meanwhile, would have the distinct function of providing the projectiles, rather like the ammunition box and belt that feed the rounds into a machine gun. What the two scientists needed to do was to create a prolific source of protons and then find a way of delivering the flow to the desired point on their finished apparatus. The obvious place to find protons in large numbers was hydrogen gas, since a proton and a hydrogen nucleus are identical, and so the challenge was to identify the best way of stripping off the electrons orbiting around those nuclei. This had to be done on a much larger scale than had previously been accomplished. Already in 1929 they had investigated this and were particularly attracted by a method developed by an American, Henry Barton, which involved diffusing hydrogen through a hot platinum tube. After long labour, however, they were forced to concede that this would never produce anything like the flow they needed and so in March 1930 they fell back on an electrical method.

In a hydrogen atom the link between the one-proton nucleus and its single orbiting electron can be fairly easily broken. If you aim an electrical current, which is a fast-moving stream of electrons, through a chamber containing hydrogen gas, the effect will be to ionize many of the gas atoms, in other words to batter them in such a way that they divide into their component parts. These broken-off electrons and protons may then be drawn off in opposite directions by taking advantage of their distinct electrical characters. This can be achieved at just fifteen volts but Cockcroft and Walton, requiring as they did a very intense flow of protons, decided to use 40,000 volts applied to a small container continuously fed with hydrogen from a separate tank. This flow they fed in turn into a pipe, or 'canal', just 1.5 mm in diameter, which in due course would direct the protons to where they were wanted. Once complete it was a neat enough piece of apparatus, following principles well known in the Cavendish, and there was little difficulty in making it work in isolation. When it came to incorporating it into the overall assembly, though, matters became more complicated. A second transformer was needed to produce the 40,000 volts, but could it be operated so close to the 300,000-volt one without disastrous interference? The answer was elaborate insulation: the smaller transformer was installed on a pillar of insulating material about a metre and a half high and protected by a large sheath of galvanized iron. In an almost comic addition, the primary power was fed to this from a little motor on the ground by the simple method of turning a loop of cotton rope – a sort of non-conducting bicycle chain.

Now at last came the acceleration tube, in which high voltage and protons would be brought together. In broad principle this was like an artillery piece, with a breech in which the protons would be subjected to the huge repulsive force produced by the high voltage and a barrel down which they would be blasted towards their target. In practice it was a glass vessel similar in size and shape to the tubes of the rectifiers, about a metre long and bulging in the middle. It stood upright and inside it at the top was its own anode, a positive electrical terminal connected to the full high-voltage supply. At the foot was a cathode, a negative terminal also connected to the high voltage, and the whole vessel was emptied of air by another of Bill Burch's Apiezon pumps. The aim was to create between the live anode and cathode the greatest possible 'potential difference', as the engineers call it, and then to hold it steady, with no actual flow of electricity – like flattening the accelerator in a racing car but keeping the clutch just above the point of jolting contact. With the tube in this highly pent-up state the scientists would then introduce their protons as close as possible to the anode at the top. Since like charges repel each other, the instant the positively charged protons encountered a positive anode charged to 300,000 volts the force of repulsion would take violent effect and the protons would be hurled away. They would not, however, fly off in random directions for there was the power of the cathode to reckon with as well. The cathode was at the other end of the tube and it carried a negative charge of 300,000 volts: since unlike charges attract each other the positive projectiles would be drawn straight towards it and in this way a dense, high-speed particle beam would be created along the length of the tube.

Two practical difficulties were obvious: how to bring the protons and the high-voltage anode together at the top of the tube and how to draw off the accelerated protons at the bottom without them crashing into the cathode and losing their speed. Both were resolved by the simple means of making the two electrical terminals, the anode and cathode, in the shape of pipes. They were like lengths of very shiny drainpipe inside the glass tube, almost meeting in the middle. Set above the anode pipe, on top of the whole apparatus, was the proton supply with its canal leading down into the live pipe. Once the particles reached that point, so the plan went, they would be seized by the repulsive force and driven powerfully downwards through the anode and out, directly into the mouth of the cathode pipe. There they would continue their downward journey until they passed out of the bottom of the cathode, by which time they would have felt the full effect of the voltage and their speed could be measured in thousands of kilometres per second – the higher the voltage, the higher the speed. After they emerged at the bottom they would be allowed to carry on into a special chamber directly beneath, where they would strike a target of gold, carbon, nitrogen or whatever other material was chosen, while the scientists observed the effects. This experimental chamber was designed so that the targets could be inserted and detection equipment attached according to need, and it was the only part of the apparatus where man and machine would come into direct contact. It was here, at last, that safety became an issue.

In general the approach to safety at the Cavendish was what the scientists would probably have characterized as a common-sense one for its time. So far as radioactivity was concerned they were aware that in direct contact with the skin radium and its products could cause burns and if the contact was sustained over a long period the damage would be lasting. The rule was that gloves should be worn, and anyone inclined to take risks had only to look at the hands of Rutherford's assistant, Crowe, for a warning. Crowe routinely prepared sources for experiments and, no doubt because the work required a delicate touch, had long been in the habit of using bare hands. The result was that he lost some of the feeling at the tips of his fingers, then the skin hardened and finally lesions opened up that refused to heal. Though he had several skin grafts one finger had eventually to be amputated. Not surprisingly, direct contact of this kind became the chief safety concern in the lab, while the more general danger from radiation in the air and on secondary surfaces was treated fairly lightly – it is indicative that the reason for keeping the stock of radium in the highest room in the building was not concern for health but a hope that any escape of radioactive material would travel up and out, rather than descending into the workrooms and spoiling any particle counts in progress. It was not until the late 1930s that the first health physics specialists made their appearance at the Cavendish and of course they found worrying levels of contamination in many rooms.

On Allibone's evidence Rutherford was much more anxious about the high-voltage equipment, with its risks of electrocution, than about his trusty radium. Cockcroft and Walton were sensible enough to keep their distance from the live apparatus so far as possible but when it came to observing experiments they knew they would have to get close. Accordingly they built a little wooden observation hut attached to the experimental chamber, where one of them could sit on a stool and observe any effects through a microscope. This they were careful to insulate thoroughly from the high voltage.

Less obvious and less well understood were the dangers from incidental radiation caused by their machine, notably X-rays, and here they provided themselves with only the most basic protection. The hut was lined with scrap lead and inside it a zinc sulphide screen was propped up to act as a monitor – 'If it glowed we added more lead,' Cockcroft would recall later. As for the general levels of radiation in the room there were no precautions; their control switches, for example, were just a few feet from the live assembly and in direct view, an arrangement which Cockcroft, looking back from the 1950s, acknowledged would be 'horrifying to modern health physicists'.[3]

By May 1930, after eighteen long months of preparation, all the main pieces were in place. The transformer was functioning satisfactorily; the rectifiers were able to cope, if not quite with the full voltage, then with something approaching it; the proton supply was up and running; the acceleration tube was ready; the various pumps and supporting electrical equipment had been installed and tested. The moment had come to put it all to work.

The machine took time to start up – care was needed in clearing the vacua, warming up the rectifiers and raising the voltage. The exact position of the proton canal often had to be adjusted and the supply of particles through it gradually increased until a green fluorescence announced a good flow. Then the voltage could be pushed up in slow stages and the results monitored. It worked at 50,000 volts. It worked at 100,000 volts. Some adjustments were made. It worked at 125,000 volts, and at 150,000 volts. And so, step by step, they climbed, alert to any signs of stress and anxious not to push things too far too soon. Every now and then they would test the beam to check whether it was picking up electrons or other contamination in the tube. A little perhaps, but not enough to be a problem. Now they reached 200,000 volts, and 250,000, and still it worked, though there were signs that it would not go much higher. At 280,000 volts they stopped, feeling they had reached the limit for the moment. The build-up to this point probably took a week or more but the tubes withstood the stress, the corona shields were doing their job and there had been no dramatic breakdown. With more work they might be able to push the voltage higher but for the moment they had made a good beginning.

The next concern was focus. It was all very well producing fast protons but little would be gained if they sprayed outwards as they descended through the tube, like pellets from a shot-gun. To be most effective the beam should be both dense and fine and it was probably no surprise to either Cockcroft or Walton, or indeed to Rutherford, that at this first attempt it was neither. A little screen was placed in the experimental chamber and the shape of the beam landing on it could be clearly seen. At 80,000 volts it was nicely concentrated in a distinctive star shape but as the voltage went higher this grew fuzzier until eventually there was just a blurred circle four centimetres in diameter. It was a problem but again it was one that could be worked on; without doubt ways could be found of improving the focus although it would involve internal adjustments to the experimental tube – which in turn would mean taking it to pieces. Before they did that, however, they intended to find out whether they could achieve anything interesting with the beam they had; having devoted so much time largely to problems of electrical engineering they were very keen to try some atomic physics. Even at 280,000 volts it was surely worth making the attempt, and Rutherford, desperate as he always was for progress, would have expected nothing less.

The question now arose: what should they look for? The

theory of the time gave very little guide as to what was likely to happen when a proton entered a nucleus, particularly a proton of relatively low energy. It seemed unlikely that it would have the power to knock out another proton so the best guess was that it would enter the nucleus and come to rest there, perhaps a little uneasily, this uneasiness expressing itself in an emission from the nucleus, not of another proton or an electron, but of gamma rays. Some curious gamma-ray effects had already been observed around the nucleus by other experimenters so it seemed worthwhile to use the new apparatus to investigate this. Accordingly Cockcroft and Walton attached to their experimental chamber a gold-leaf electroscope, a venerable but reliable indicator of the presence of gamma rays. Then they placed a target in the line of their beam. At first they used a small sample of the element lithium, and found nothing. Even taking the voltage right up to 280,000 volts brought no response whatever from the electroscope. So they tried a different target, beryllium, and this time the detector responded: some form of radiation was emerging, although it was very, very faint. Like lithium, beryllium was a light element, so just to see what would happen they tried something much heavier under the beam: lead. Once again at fairly high voltages there was radiation although it too was barely detectable. Were these the gamma rays they hoped for, coming from a nuclear reaction? Or were they caused by some unwanted effect in the acceleration tube? Why would these rays emerge from experiments with beryllium and lead and not lithium? This was the sort of work they were trained to do and had set out to do. Just as they got down to it, however, their apparatus let them down. First the focus began to deteriorate, although with some effort they were able to correct this. Then the transformer failed, which was far more serious. It looked like a setback of several weeks, but as things turned out they were never to complete the experiments or establish the character of those strange rays.

In several respects this whole story pivots on the summer and autumn of 1930, when significant developments took place in the world of physics, in the Cavendish and, not least, in the lives of the occupants of that basement room. For Walton in particular it was a turning point. For one thing his scholarship ran out and there was no possibility under the 1851 statutes of extending it into a fourth year, so once again he found himself in a position that would have unsettled a less phlegmatic man. After three years at the laboratory he still had no Ph.D. and only a single published paper to show for all his efforts; by comparison with his contemporaries, who were bringing out two or three papers a year and beginning to land Fellowships in Cambridge colleges, he was almost invisible to the scientific and academic world. In the circumstances it is easy to see why he was careful to nurture whatever contacts he had that might lead to future employment. He kept in touch with Trinity College, Dublin but, though there were signs that his old physics department was being shaken up, no vacancies had yet emerged. He had also been stung a year or so earlier by the college's refusal to allow him even to sit the examination for a Fellowship. It is clear that he also showed some interest in working at Metropolitan-Vickers, where those in charge of the research department were well disposed towards him. Here too there were difficulties, though, for the company was feeling the first cold winds of the Depression and in the uncertainty that prevailed no job offer could be made.

It was Rutherford's patronage that rescued Walton, just as it had a year earlier. Ever since the war, through its new Department of Scientific and Industrial Research (DSIR), the government had been increasingly drawn into support for university science, and it was natural that Rutherford was an influential figure in the relevant committees. By the same token, given the standing of his laboratory, it is hardly surprising that a good many grants from the DSIR went to researchers at the Cavendish. In June 1930, therefore, he proposed to the scientific grants committee (of which he was not a member) that they should give Walton a senior research award. The reference he wrote is worth quoting:

He [Walton] has held an 1851 Exhibition Scholarship for three years and in our opinion is a man of exceptional ability who has acquired the very difficult technique of work in high vacua with high voltages. This has involved a great deal of work of trial and error, since very little information is as yet available in this direction. For this reason he has not had the same opportunity to publish papers as a man of equal ability who is engaged on simpler and more direct problems. Quite apart from the ability and promise of Walton, I feel it of great importance that the work on which he is engaged with Dr Cockcroft, viz. the production of high speed atoms, should be developed as rapidly as possible, and Walton is the only student we have who has the ability and the technique to carry it out.[4]

The committee duly approved the award.

While this gave Walton some security for the next three years it would be too much to say that it made him comfortable. The DSIR paid £275 for the first year, which was actually £5 less than his 1851 grant, and it must have been a struggle to get by in Cambridge on that income. We have a good idea of the cost of living there thanks to a survey carried out by the 1851 Commission among the students it supported in 1929: board and lodging ate up about £140 a year for a single person and college fees of various kinds another £65, leaving the 1851 student with the grand sum of £75 to pay for a year's clothing, laundry, books, stationery, tools, medical expenses, routine travel and other needs (bicycle parts were a common requirement). For many this was not enough. George Laurence, who responded to the survey in spirited terms, calculated that he had spent £346 in the year 1928–9, presumably relying on savings or family for the £66 excess, and he was adamant that no high living had been involved. Life in Cambridge, he declared, was needlessly expensive because the rents for college-approved lodgings were set too high and prices in the town were generally inflated by the prodigal habits of well-off undergraduates. Research students were the poor relations of the university, and for this he blamed an establishment of 'classics and literary men' who had lumbered them with 'regulations so suggestive of a monastic order in the twelfth century'.[5]

How Walton got by is not clear. He had £200 in savings in 1929 but the income of around £10 a year can hardly have made a great difference, while it is doubtful that his parents were able to provide much support. The likeliest explanation is thrift – eating all his evening meals in college, where the fare was so hearty, skimping on lunches and passing his holidays cheaply at home. (Laurence, who was from Canada, was unable to do this last and was naturally inclined to see something of Europe while he had the chance. He argued: 'The £50 spent in vacations during the twelve weeks that the laboratory is practically closed to us might be challenged on the ground that £20 of it might be saved by remaining in England. Personally I consider the £20 well invested.')[6]

However he coped, Walton made no complaint about his income and was no doubt happy in 1930 merely to be allowed to remain at the Cavendish. With the apparatus complete, moreover, he was at last able to do something about his meagre publishing record; in August he and Cockcroft sent an account of their joint work to the Royal Society. 'Experiments with High Velocity Positive Ions' duly appeared in the society's *Proceedings*, an event that effectively marked their public debut as research partners. The paper began with a simple statement of their objective:

It would appear to be very important to develop an additional line of attack on problems of the atomic nucleus. The greater part of our information on the structure of the nucleus has come from experiments with alpha particles and if we can supplement these with sources of positive ions [protons] accelerated by high potentials we should have an experimental weapon which would have many advantages over the alpha particle. It would, in the first place, be much greater in intensity than alpha particle sources . . . It would in addition have the advantage of being free from penetrating beta and gamma rays which are a complication in many experiments, whilst the velocity would be variable at will.[7]

Next, and for the first time, they made public Cockcroft's interpretation of the Gamow theory, suggesting that 300,000 volts should suffice to penetrate the nucleus, instead of the millions generally assumed. Of course they had no choice but to reveal this since otherwise their whole project would appear foolish, but at the same time they must have known that by doing so they were alerting other laboratories to the possibility of a short-cut to the nucleus; from now on, in other words, the risk of competition increased sharply. The rest of the paper was a straightforward account of the apparatus, with observations on the choices that lay behind it and the difficulties encountered. It concluded with a brief reference to experiments 'of a preliminary character' and to the 'very definite indications of a radiation of non-homogeneous

Experiments with High Velocity Positive Ions.

By J. D. COCKCROFT, Fellow of St. John's College, Cambridge, and E. T. S. WALTON, 1851 Overseas Student.

(Communicated by Sir Ernest Rutherford, P.R.S.—Received August 19, 1930.)

[PLATE 21A.]

1. *Introduction.*

It would appear to be very important to develop an additional line of attack on problems of the atomic nucleus. The greater part of our information on the structure of the nucleus has come from experiments with α-particles and if we can supplement these with sources of positive ions accelerated by high potentials we should have an experimental weapon which would have many advantages over the α-particle. It would, in the first place, be much greater

From the *Proceedings of the Royal Society*:
their first joint paper

type' they had found on bombarding beryllium and lead. No reference was made to the failure of the transformer.

The article, in fact, leant heavily on the final report Walton had written a few weeks earlier for the 1851 Commissioners, which suggests that, just as he was responsible for most of the toil in the laboratory, in this case the Irishman did most of the writing-up as well. This accords with what we know of Cockcroft's life at the time, for in June 1930 he and Elizabeth suffered further tragedy with the birth of a stillborn child. The impact on the couple at home can only be imagined: they had waited nine years to marry and had been together five, and now they had lost two children in seven months. Cockcroft's response at the laboratory was evident to colleagues and friends: he buried himself still deeper in his many jobs. As it happened, his responsibilities in Kapitza's magnetic laboratory were increasing rapidly at this time because an important transition was under way there. Hitherto the enormous cost of this project – which reached £2,750 in 1930 – had been met by the DSIR, but negotiations were under way to transfer the burden from a government increasingly squeezed by the Depression to the Royal Society. Rutherford's plan was to create a whole new building for his Russian protégé and fill it with a new generation of magnetic equipment. The chosen site was the only one available, the Cavendish courtyard, which would mean demolishing the existing workshops. It was the director's most ambitious undertaking since his arrival in 1919 and Cockcroft was intimately involved in the planning and negotiations. The deal was finally done at the end of the year, when the Royal Society took over the payment of Kapitza's salary and pledged a breathtaking £15,000 grant for the new laboratory. Building would start the following year and the demands on Cockcroft grew accordingly.

Walton, meanwhile, having apparently settled his future for the next three years and at last reached the point of publishing something of substance, was passing another milestone and it was a happy one, for that July a chance encounter at a railway station brought romance into his life. It was the summer holiday and he was in Ireland. His existence there had always been peripatetic: born in Dungarvan on the south coast, he had lived in no fewer than seven other towns since then, most of them in the north, as his father moved from parish to parish. Working in Ulster had caused John Walton some discomfort over the years because he was a Home Ruler while most northern Methodists were not – back in 1912, for example, he had annoyed some of his parishioners in Cookstown by refusing them permission to place their 'Solemn League and Covenant' at the church door, to be signed by worshippers loyal to the Crown. By 1930 Ireland was partitioned and John Walton was still serving an Ulster parish, this time in Coleraine near the north coast, so it was there that Ernest passed his summer holidays.

His habit was to pay visits while in Ireland. His mother's people lived in County Armagh and he usually called on them; though she had died when he was still a toddler (his father had remarried) the connection with her family, the Sintons, was a strong one. Walton also liked to spend time in Dublin, where his sister Dorrie lived and where he could refresh his connections at Trinity College. It was quite normal, therefore, that he should have travelled on the train between Dublin and Belfast a few times in the course of his holiday and it was on one such journey that he had his unexpected meeting. The train developed a fault and came to an unscheduled halt at a country junction called Goraghwood, where all the passengers were required to alight and wait on the platform for another train to take them to their destination. In the crowd Walton recognized a face from his schooldays at Methodist College in Belfast: Winifred Wilson. They fell into conversation and, almost as swiftly it seems, they fell in love.

Like Ernest, Winifred was a child of the manse, her father at that time serving the parish of Donaghadee, east of Belfast. After leaving school she had trained as a kindergarten teacher and in 1930 she was running the infants department of Bishop Foy's School in Waterford on the south coast of Ireland, making her a fairly frequent traveller on the Dublin–Belfast line. The two had been sweethearts at school but they did not keep in touch when Ernest went off to university. Now, meeting for only the second time in eight years, they chatted happily on the platform, continued their conversation when the journey resumed and exchanged addresses before parting at their destination.

Winifred had mentioned that she planned to visit London in August to attend a course and Ernest was quick to write encouraging her to come and spend a day in Cambridge. Although the letter can have left her in no doubt about his interest it was businesslike in tone, beginning 'Dear Miss Wilson' and including none of those happy phrases a young woman might savour. This man was a scientist, not a poet. She had the measure of him, however, joking in her reply about addressing an old school friend as 'Mr Walton'. Yes indeed, she said, she would have some free time when she was in London and she suggested two or three possible dates. The reply was swift, though again it began 'Dear Miss Wilson': would she like a trip on the river Cam on the Saturday? Straining to add some small talk, Walton went on to report that his parents had recently paid him a visit and that he had been working hard in the lab by day and writing 'some reports on my work' in the evenings. Then, worried he was talking shop and fearing he might not be capable of much else, he warned: 'You might be bored stiff with me after a day.'[8] She took the risk, the visit took place and a letter soon afterwards announced that she had had a splendid time. 'I hope your results will prove successful so that you may get your holidays soon,' she said, urging him to come and visit her in Donaghadee on his way home to Coleraine.[9]

That second meeting did not come so quickly, Walton showing a manly carelessness about the arrangements. Her visit to Cambridge was made during the Cavendish's long vacation term and the holidays that she referred to were the break that followed, from mid-August to late September. Her own school term, however, began in early September and took her south to the other end of Ireland. Walton not only missed his opportunity to see her before she left Donaghadee but also failed to write. Winifred, not content to let things lapse, took up her pen again at her lodgings in Waterford to express mild anxiety at his silence and report chattily on the three rooms of 'very lively youngsters' she was now teaching.[10] His chastened reply declared that he was always an unreliable letter writer who could be counted upon to communicate with his own parents only about once a month. Though she was now writing 'Dear Stew' and signing herself 'yours sincerely, Freda Wilson', Walton remained 'Ernest Walton' and she was still 'Dear Miss Wilson' to him.

Despite the stubborn formality, the phase of stumbling preliminaries was coming to an end and the correspondence assumed a steadier pattern. It was to continue until their marriage four years later, with exchanges each week and sometimes oftener. The two had much in common, not only a background in the manse and shared memories of boarding school but also a religious outlook that included a commitment to hard work and disapproval of alcohol and gambling. Neither smoked, though Ernest had tried it – he was 'not sufficiently impressed to be induced to make a habit of it'[11] – nor did they dance. Despite these 'good Methodist views'[12] they were not in the least dour, and nor, with rare exceptions, were they inclined to judge others who had different standards. Freda emerges from the letters as a warm, lively young woman and by force of encouragement and example she manages slowly to breathe some warmth into Ernest's letters. After four months she at last became 'Dear Freda', part of a compact under which she dropped 'Dear Stew' – which he disliked – in favour of 'Dear Ernest'. Though it would be too much to say that he ever acquired a romantic touch he learned to include some human content as well as the occasional observation on life in Cambridge. 'I can always blame my scientific upbringing for any shortcomings in my letters,' he wrote, 'for here I am continually trying to express facts and facts only in as concise a manner as possible.'[13]

Another characteristic the couple shared was discretion, and by agreement they concealed their growing friendship from others, including Ernest's sister Dorrie in Dublin, whom Freda knew well from schooldays. Nor do they appear to have acknowledged openly to each other that this was a romance, although that is what it was. At New Year, after an exchange of Christmas cards (his arrived late), they met again. He was travelling back through Belfast to catch the night ferry to England and she came up from Donaghadee for a rendezvous at 4.30 p.m. outside Anderson & McCauley's department store on Royal Avenue. It was the first of several meetings at the same spot.

On the scientific front developments were no less significant, for in the course of that busy summer and autumn Cockcroft and Walton came to a decision which seems at first surprising, even astonishing: instead of pressing on with their gamma-ray experiments once the transformer fault was repaired, they chose to rebuild their whole machine almost from scratch in order to more than double its capability. An 800,000-volt assembly would replace the 300,000-volt one. The ramifications were considerable, for this would not only cost money but also require a whole new round of the sort of construction work that had occupied them for the past year and a half. And it would cost time, for until the new apparatus was complete, which would be several months at least, there would be limited opportunity to conduct nuclear experiments. This, at the moment when they had just revealed to the scientific world their view of the possibilities opened up by the Gamow theory; for all they knew, someone in an American or European lab might seize that opportunity immediately.

There were several reasons behind the decision and together they amount to a strong case. The first and most compelling was that their existing apparatus, though it might be capable in time of producing a proton beam at their original target of 300,000 volts, would always struggle to do so. The lesson of the transformer failure was that at maximum output breakdowns would be frequent and sustained experimental work difficult if not impossible. And 300,000 volts was in practice the *minimum* level at which Cockcroft's calculations suggested that results of interest were likely. What they had, in short, was at its best barely enough to do the job, and its best was impossible to rely upon. Behind this practical consideration lay something more profound, for it seems certain that neither Cockcroft nor Walton really trusted that figure of 300,000 volts. They knew that Gamow's calculations depended upon a number of approximations, as in turn did Cockcroft's, so the idea that the workable minimum energy for penetrating the nucleus might not be 300,000 volts but 500,000 or 700,000 or even 1 million would not have surprised them. It would take only a small adjustment to the equations to produce such a variation. In fact their article in the *Proceedings of the Royal Society* that year leaves the impression that it was always their intention to build a second and bigger apparatus. The 300,000-volt level had been chosen, they wrote, 'because it appeared quite possible to begin work on a laboratory scale with voltages of this order . . .' and there was 'at any rate some chance of obtaining results of interest'. The language has a preliminary sound.

Rutherford, for his part, always had doubts about the theory and had given his support to Cockcroft and Walton as much in the hope that this was a first step down the high-voltage route as that it would actually produce nuclear disintegrations of its own. He made these views clear in 1930 in a speech he gave when opening a new high-voltage laboratory at Metropolitan-Vickers in Manchester. Not once did he allude to the possibilities opened up by Gamow; instead he looked forward to million-volt university laboratories in much the same terms as he had when talking to the Royal Society in 1927:

There is one aspect of this high potential problem that specially appeals to me. I refer to the application of very high potentials to highly exhausted tubes in order to obtain a copious supply of swiftly moving electrons and charged atoms. As far as I am aware the highest potential so far applied to a single discharge tube is about one million volts and this is a difficult technical problem. In these respects Nature far outstrips our puny experiments in the laboratory . . . What we require is an apparatus to give us a potential of the order of ten million volts which can be safely accommodated in a laboratory. We require too an exhausted tube capable of withstanding this voltage.[14]

Cockcroft, Walton and Allibone were all present, as were senior Metro-Vick scientists such as Fleming and McKerrow, who had been supporting the Cavendish accelerator project from a distance. All must have been surprised by such words and yet Rutherford was only expressing the orthodox view. For anyone familiar with the frustrating rarity of disintegrations by natural alpha particles travelling at speeds which, if they were artificially achieved, would require something like 7 million volts, the idea that by some trick of Russian theory protons at 300,000 volts might do the same job could be hard to swallow.

Events in America also conspired to push the Cavendish men towards higher voltages. In September 1930 Charles Lauritsen published an account of his second-generation apparatus at Caltech in Pasadena, and it was a veritable Rolls-Royce of a machine. With the potential to operate at 1 million volts, his tall tube of petrol-pump cylinders now stood in an iron frame rather than a wooden one and its base and control desk were housed in concrete compartments to give protection. Many of the problems of corona effect appeared to have been conquered and the drawings in the *Physical Review* gave the impression of a finished product rather than a test bed. Lauritsen's interest was still focused on X-rays rather than nuclear effects and his apparatus continued to be tied up treating patients, but if ever he chose to attempt nuclear bombardment experiments it could probably be adapted in a few weeks. In Washington, meanwhile, Merle Tuve and his colleagues were also pressing ahead, with notes in the published journals recording first that the tube in the oil tank could withstand 1.9 million volts and then that powerful streams of electrons were being measured. From Germany too came news that Brasch and Lange, working with the AEG voltage source, had successfully applied 2.4 million volts to their massive porcelain tube. And a further spur to action came when reports reached Cambridge of strange developments at Berkeley. Until this time the California laboratory had barely registered as a place with an interest in atomic matters, but now Ernest Lawrence announced to a scientific gathering that he and a colleague had built a prototype circular proton accelerator – and that it worked. 'Preliminary experiments indicate,' Lawrence stated boldly, 'that there are probably no serious difficulties in the way of obtaining protons having high enough speeds to be useful for studies of atomic nuclei.'[15] This was remarkable enough to merit coverage in the *New York Times*.

It is easy to imagine Rutherford surveying these developments and concluding that it was a waste of time for his team to be fiddling about with 300,000 volts. The sooner they made a start on something more ambitious the better, even if for the moment the million-volt threshold seemed beyond their reach. And the slender results they had achieved with their first bombardment experiments seemed only to confirm that much more power was needed if something worthwhile was to be done. Besides all this, there was one final, simple force pushing Cockcroft and Walton towards higher voltages and that was the belief that they could do it. On the practical front they now had the space to expand their equipment because in the summer Allibone had left to become director of the new Metro-Vick high-voltage lab (a post Cockcroft turned down). And more importantly, they had found a way of pushing their voltage higher without having to buy another expensive transformer. So it was that, when Cockcroft and Walton returned for the new academic year in October 1930, their minds were focused on building an 800,000-volt accelerator. It is one of the teasing ironies of this story that, however logical this step seemed at the time, and however pressing the case for it appeared to be, in scientific terms it was not in fact necessary.

11. Off to the Races

In late 1930 the small international community of experimental atomic physicists was experiencing a new optimism
and vitality. For them there was work to do and it promised
at last to make possible an escape from the dead end they had
found themselves in at the close of the 1920s. No such mood,
however, existed among the theoreticians. When Gregory
Breit toured the continental homeland of theory late in 1928
he had been struck by the 'comparative sterility' of debate on
atomic matters and three years later things had got worse
rather than better.[1] In fact theorists probably knew even less
about the constitution of atomic nuclei in 1931 than they had
in 1928. This paradoxical change is well illustrated in the
activities of George Gamow.

We last saw the Russian in early 1929, making his way back
to Copenhagen from England having visited the Cavendish
and spoken at the Royal Society. His thoughts were already
moving on from the possibilities of particle bombardment to
a more general view of the nucleus founded on quantum
mechanics. He saw it as resembling a water droplet, with
a skin formed by a kind of surface tension – an idea that
would be taken up and developed by Bohr, among others.
Gamow remained in Copenhagen working on this and
entertaining his colleagues until his Carlsberg award ran out
in the summer, by which time he had secured a Rockefeller
fellowship to take him back to the Cavendish for a year. His
references could hardly have been better: Rutherford and
Fowler both sang his praises to the assessors while Bohr stated

that he regarded the young Russian as 'another Heisenberg'.[2]

In the autumn, therefore, Cambridge once again experienced the diverting hubbub of a Gamow arrival. This time he chose to fly to England and he informed friends on reaching the lab that he had thrown letters addressed to himself out of the aeroplane window and was hoping they would find their way back to him. There was further amusement when he found lodgings in a house called, of all things, 'Kremlin', and more followed when he bought himself a BSA motorcycle. The Cavendish had experience of Russians on motorbikes since Kapitza also bought one in his first days at the lab – both he and Chadwick nearly killed themselves on it. Gamow seems to have been more prudent but he still managed to outrage Rutherford at least once. Bohr paid a visit to Cambridge that year and during afternoon tea at the Rutherfords' one Sunday Gamow (evidently unaffected by Lady Mary's notions of proper behaviour) encouraged the Dane to take his BSA for a spin. No sooner was the great man out on the public highway than he stalled and skidded, bringing traffic to a halt, and Gamow recorded the sequel in verse:

> While Gamow rushing to the fore,
> Was doing what he could for Bohr
> Who should like Jove himself appear
> But Rutherford? In Gamow's ear
> He thundered: 'Gamow! If once more
> You give that buggy to Niels Bohr
> To snarl up traffic with, or wreck,
> I swear I'll wring your bloody neck![3]

Although he spent a full year at the Cavendish there is no sign that Gamow exerted any further influence on the work of Cockcroft and Walton. He became friendly with Cockcroft, who made a vain attempt to teach him golf (Gamow spent some of his Rockefeller money on a pair of plus fours, but they did nothing for his game), but so far as the science of the accelerator project was concerned it appears there was little more he could contribute. An important event in his stay, however, was a commission from the Oxford University Press to write what would be the first book published anywhere about the theoretical physics of the nucleus. The idea appears to have come from a former Cavendish student called J. G. Crowther, who among other things was a scout for the Oxford publishers and who persuaded Fowler and Kapitza to contribute to the same series.

In the case of the Gamow book the path to publication was a rocky one, for while he was clearly the best person in Europe to tackle the subject his idiosyncrasies could not be kept under control. The text had not been written by the time his stay in England ended in summer 1930 and so the task accompanied him back to Copenhagen, where it had to compete for his time with table tennis, parlour games and amateur dramatics. Dirac, who was at the Bohr institute that autumn, later told Crowther the book was largely written on the ping-pong table, between games. The real problems began, however, when the manuscript finally reached England, for it was barely comprehensible, not because the science or thinking were obscure but because of the chaotic jumble of languages in which it was written, German and French spellings jostling with English and Danish vocabulary on a background of Russian syntax. Fowler balked at the job of translation and passed the text to a former research student of his, Bertha Swirles. It was a mighty labour – 'there is an occasional correct sentence',[4] she observed, looking on the bright side – and even when she had finished she felt the need to pass it to a second colleague to sift out what she called 'residual Gamow'.[5]

Unusual though all this may have been it was only to be expected, indeed it is hard to believe that Crowther and the OUP were not warned. Far more important than the language, however, was a problem that had arisen in the subject matter itself, a problem which greatly troubled Gamow.

The nub of it was the old idea that electrons must exist inside the nucleus, that simple notion which could be traced all the way back to the need to account for the difference between the weights of nuclei and their electrical charges. Only Rutherford had ever come up with an alternative theory, the neutron, and he had been unable to produce any evidence for it. Gamow therefore had little choice but to repeat the long-standing orthodoxy. 'We assume,' he wrote bluntly (once he had been translated), 'that all nuclei are built up of elementary particles – protons and electrons.'[6] And yet in recent years the theorists had become increasingly uncomfortable about this whole business. Quantum mechanics had vastly improved their understanding of electrons and by now they simply could not imagine how such creatures might exist inside nuclei, where they must be locked in close proximity with protons. Gamow expressed things diplomatically when he said there was a striking disagreement between experimental evidence and theory – so striking, in fact, that if there had been no evidence to the contrary theorists would have taken it that a nuclear electron was impossible. And yet the evidence to the contrary was really very straightforward. In the process of radioactive decay electrons emerged from the nuclei of radium, uranium and other elements, and if they came out, then surely it was reasonable to say they must have been inside in the first place? But was it so reasonable? In logical terms the assumption was not quite so airtight as it seemed and the doubts were well illustrated by a question raised at the time: does a gun contain smoke? Certainly smoke emerges from a gun when it is fired (just as the electron emerges from the radium nucleus), but that does not mean that you would find smoke inside a gun if you took it apart. It was not a very scientific analogy but it made the point: there was no hard evidence that electrons existed inside nuclei.

So uncertain had matters become that Gamow decided drastic measures were needed in his book. He bought a rubber stamp with a skull and crossbones and wherever a reference to the nuclear electron appeared in his text he added the symbol in the margin. He told his publishers he wanted them to reproduce these marks in the book so that readers would be alerted to the questionable character of the content. It was too much for the OUP – you did not need to be stuffy to believe that pirate flags had no place in a scientific monograph – and they dug in their heels. In due course a compromise was reached under which the offending symbol was replaced by a bold wavy line that was something between a Spanish tilde and a snake. 'These signs,' announced Gamow's demure note on page two, 'mark the more speculative passages about nuclear electrons.' The theoretical vacuum in nuclear studies was not easing but growing more problematic and it is little wonder that Gamow politely pressed Cockcroft to get on with disintegrating atoms and that Bohr wrote to Rutherford of his 'longing for new experimental facts'.[7]

The first physicist to produce new facts, however, was not a high-voltage specialist nor even a Cavendish researcher. In August 1930 Walther Bothe and his assistant Herbert Becker published a short note in *Die Naturwissenschaften*, the weekly journal that was Germany's equivalent of *Nature*. Following the new experimental approach discussed with Chadwick and others as far back as 1928, these two had been quick to bombard light elements with alpha particles from polonium and to record the results with a Geiger counter rather than a zinc sulphide screen. The technique was successful and their first findings were surprising: in certain cases the counter had registered the presence of some very powerful rays. A few months later the Berlin physicists reported further progress: reactions were occurring in lithium, beryllium, boron, magnesium and aluminium and the rays from beryllium in particular seemed to be astonishingly strong. They were definitely not alpha particles or protons, but appeared to be gamma rays. No one could explain such a reaction for this was not the familiar nuclear snooker revealed by Rutherford in his disintegration experiments, in which one particle entered the nucleus and knocked another out. Gamma rays are like light and although like light they can behave like particles these are not remotely on the same scale as alpha particles; a release of gamma rays was more like an overflowing of energy than the click–click of snooker balls.

This discovery gave both theorists and experimenters something to get their teeth into at last and Chadwick in particular was intrigued. Somehow he had managed to add a little to that small supply of polonium sent to him by Lise Meitner and for about a year a research student called Hugh Webster had been struggling to achieve results with it. The source was still too weak, however, so that just as Bothe and Becker published their findings Webster was completing a disappointingly inconclusive Ph.D. thesis. Now Chadwick threw him back into the chase, telling him to repeat the German work on beryllium, and soon the young man had teased out an intriguing fact about the new reaction. The new radiation appeared to travel most strongly in the same direction as the alpha particle which provoked it; in other words, it travelled onward from the collision point and not sideways or backward. As Chadwick said later, this discovery 'excited me very much indeed, because I thought, "Here's the neutron." '[8] Perhaps this really was nuclear snooker; perhaps the rays that Bothe and Becker had observed were not lightweight gamma rays after all, but big neutral particles knocked out of the beryllium nuclei. Once again, however, Chadwick was disappointed because the neutron, if it was present in these rays, was still too well camouflaged for him to detect. The students at the Cavendish produced a sketch that year about the 'fewtron', a particle so elusive you could tell it was there only when nothing at all showed up on your detection equipment, but Chadwick was not laughing. It was plainer than ever to him that if the Cavendish were to get ahead of the Germans again, and if he was to be the one to find the neutron, he needed more polonium. It so happened that a supply was about to present itself.

Chadwick's diffidence was familiar to all the students. Allibone was only one of several to describe taking a problem to the assistant director, explaining it at length and then leaving his office without any confidence that the man had even been listening. Sometimes he would say a crisp 'yes', but without indicating whether this signified agreement or merely that he had heard what was said. This lack of engagement could be frustrating, and it could be amusing. Later in life Chadwick would run a department of his own at Liverpool University and a story was told of him descending to a workroom there in search of a particular student, whom he found helpless on the floor with the edges of his overalls neatly nailed to the boards beneath. Chadwick leant over him, elicited the answer to the question he had come to ask and withdrew. Only when he had regained his office did he mutter casually to his secretary that 'the boys have been up to their tricks again'.[9] It is easy to see why people in the outer orbit of his acquaintance found him distant and forbidding and yet the experience of the young physicists who knew him better was different. Fussy and irritable he might be, but he was also a master experimenter and a clear thinker. Once they learned his ways the students who worked directly to his orders often became loyal disciples.

One such was Norman Feather, a Yorkshireman who had come up through the Cambridge undergraduate course – in fact he was the callow student who in 1926 had presented Rutherford with his menu of personal theories on nuclear constitution. Feather impressed Chadwick with some work on the distances alpha particles could travel and in 1929 landed a post at Johns Hopkins University, where he was to set in motion some radioactivity research. Arriving in the United States he looked around for some radium with which to work and found it on his doorstep at the Kelly Hospital in Baltimore. Richly endowed, the Kelly made extensive therapeutic use of radium and maintained a stock of five grams – ten times what Rutherford and all his students had to work with at the Cavendish. The hospital had no use for the radium products once they had passed a certain stage in their decay and so administrators were happy to give Feather, free of charge, some small bulbs of material they regarded as spent. This was more than adequate for experiment.

While in America Feather kept in touch with his mentor, Chadwick, and this had three consequences of interest. The first was that he was asked to visit a laboratory at Columbia University in New York which reported some findings the Cavendish physicists simply could not believe. Chadwick, suspecting a repetition of the Vienna problem, sent Feather to investigate and had his doubts confirmed. The scintillation counting at Columbia had been done by one man working for six to eight hours a day, far more than would have been permitted in the Cavendish and enough, in Feather's words, to justify saying: 'No go.'[10] The episode was revealing not only because it further underlined the threadbare character of scintillation counting but also because it illustrated – as did Feather's recruitment to Johns Hopkins – the growing American interest in atomic and nuclear studies. A second fruit of Feather's contacts with Chadwick was that they allowed him to return to Cambridge after only one year, as he had always intended, and slip straight back into the laboratory's experimental stream. And this in turn brought the third consequence. As Feather was preparing to sail for England in summer 1930 he approached his friends at Kelly Hospital to say, as he recalled later: 'Look here, in the dead bulbs that you have there is more polonium than anyone else in the world has, except perhaps the Curie-Joliots in Paris . . . What about letting me have some of this?'[11] They promptly handed over 300 of them, which he placed in an aluminium box and added to his luggage for shipment home. By this unlikely route Chadwick at last acquired the means to create a decent stock of polonium.

Down in the Cavendish basement, which they now had to themselves, Cockcroft and Walton were also very busy, although from the record it is difficult to establish the pattern of their work. It appears that while the initial experiments were under way in preparation for the 800,000-volt apparatus they continued to operate the 300,000-volt machine for some time. In October 1930 Cockcroft reported to Gamow that they had 'protons running very nicely up to 300,000 volts' and were 'studying continuous radiations', which suggests that they were still pursuing the gamma-ray traces they had detected before.[12] But at the same time Cockcroft was completing the design of an ingenious method for raising the voltage higher, and early in November they applied to the university for a grant of £100 to buy additional equipment. Then whatever experiments they were conducting were interrupted by another breakdown of the transformer. December found Walton painting a gloomy picture in a letter to Freda: 'The rate of progress in the lab has been zero lately. I spent the whole of last week looking for a very small leak in a complicated piece of apparatus. In the end I had to take it all to pieces. It is now assembled again and I hope that it will work more satisfactorily.'[13]

It must have been difficult to concentrate at this time because they had just learned of a development which had nothing to do with science but would add another substantial delay to their work. Partly as a consequence of the decision to build the new magnetic laboratory and partly because the engineering department was moving out of some neighbour-ing buildings, a general shuffle of accommodation was taking place at the Cavendish and one of the casualties was to be their basement room. It had never quite belonged to the physics people but had been on extended loan from the Department of Physical Chemistry just across the corridor and, as part of the reshuffle, it was to be returned to its rightful owners. But Cockcroft and Walton were not left homeless. Cockcroft himself was at the centre of the departmental negotiations so it was probably no accident that a decision was taken to transfer their high-voltage work to a handsome new room that was being added to Rutherford's domain. This was the former Engineering Lecture Room D in the Balfour Library, a large edifice in the tangle of buildings behind the Cavendish façade. Once the tiered wooden seating in the room was removed it would offer a far more generous and brighter space in which to build the 800,000-volt machine. In the words of Cockcroft's formal proposal, addressed to Rutherford but clearly meant for the university's buildings committee, the move 'would be very advantageous'. He went on: 'We could do all the work we required at 800,000 volts without any serious alteration of Lecture Room D. If higher voltages should prove possible and desirable in the future, structural alterations to enable more head room to be obtained would not prove difficult.'[14] He appended a list of alterations that would be necessary, including clearing the room, strengthening the floor and rewiring, and estimated a cost of around £350. All was agreed and the move was set for early summer 1931.

In the intervening months the two men made no attempt to build the 800,000-volt apparatus in the basement room but instead carried out all the preparatory work they could. Since this was largely of the kind for which Walton was best equipped, and since he was now finally writing his Ph.D. thesis as well, it was a busy time for the younger man. Cockcroft, in any case, had made his contribution by supplying the design for what came to be known as the Cockcroft-Walton voltage multiplier, a refinement of circuitry designed a dozen years before by a German engineer called Schenkel. It involved the use of an electrical component we have not met before, one which holds up or stores an electrical charge: a condenser, in essence a stack of metal discs insulated from one another, usually by waxed paper. Available commercially – they looked a little like Post Office pillar-boxes – condensers could be charged or loaded to a maximum capacity and then discharged at will, and Cockcroft's new multiplying system would employ four of them, each with a capacity of 200,000 volts. Condensers are in some ways like water vessels: if you charge one to 200,000 volts and then connect it to another that is uncharged they will split the load evenly between them, each settling on 100,000 volts.[15] Now disconnect them, top up the first one to 200,000 volts and then reconnect them and they will split the total again, both settling on 150,000 volts. Keep this process up and with a single transformer you can soon bring both condensers to capacity (or very near it). Cockcroft's design would spread this process over four condensers in a sort of cascade and arrange them in such a way that their charges would combine to provide a single power source of nearly 800,000 volts. A lot of complicated connecting and disconnecting would be required to keep all the condensers topped up in this way and it would have to be done at very high speed – many times per second. The beauty of the system as he designed it was that the timing would be provided by the alternations in the a.c. current coming from the transformer while the switching would all be done by the use of rectifiers responding to those alternations. And better still, the end product was direct current.

Elegant and ingenious on paper, in practice this was by some margin the most difficult thing they had attempted to date. The original system of two rectifiers had caused enough trouble, but now four were needed for the new arrangement and since they would have to function in such a different way their internal parts would have to be designed all over again. The prospect of all that testing and sealing and all those breakages and repairs was too much even for the ever-patient Walton and early in 1931 he and Cockcroft took another important decision. They would discard altogether the fragile rugby-ball bulb design that they had taken from Allibone and switch to something that closely resembled the apparatus of Charles Lauritsen at Caltech – the stack of petrol pump cylinders. For their new purposes this was a superior design in every way. Big, thick and robust, such cylinders would be able to withstand the voltage better and at the same time could be placed one on top of the other, the glass alone supporting the weight. With this a single tower of four rectifiers became possible, all of them hooked together and evacuated by one Burch pump. The chief drawback was that no oven would be big enough to heat such a tower as a single unit and so the whole affair could not be sealed properly with wax. Lauritsen

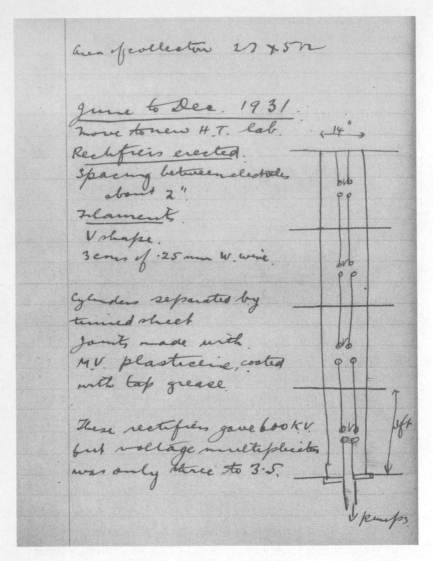

Walton's sketch for a rectifier tower
using glass cylinders

had overcome this by using steel brackets and heavy coats of shellac but this would be very troublesome when it came to repeated dismantling and reassembly. Cockcroft and Walton had a better idea: plasticine.

As the spring months passed they conducted a series of experiments with glass cylinders and plasticine under high voltages, laying the groundwork for construction of the tower when they moved into Lecture Room D in the summer. When they had time they also continued to carry out occasional nuclear experiments with their existing machine, pushing the voltage up to a peak of 290,000, bombarding lithium and looking for gamma rays, but still finding nothing worthy of note. And as if this was not enough to keep him occupied, when Walton left the laboratory at 6 p.m. it was to face long evenings of study at home, because his Ph.D. thesis had to be ready by mid-May. He reported to Freda in April that he was up until 12.30 or 1.30 a.m. every weekday night, typing and retyping a script that ran to 30,000 words. The bulk of it was a fresh account of the apparatus but he was also writing a study of galvanometers, instruments much in use in the laboratory. On Sundays, of course, he did not work, and his Sunday evenings were routinely devoted to writing his letters to Waterford.

'I'm afraid I get practically no time at present for light reading,' he declared in one of these. Even after three years at Cambridge he felt that he had 'a lot of leeway to make up' in physics because of the shortcomings of his undergraduate training in Dublin, and also there was the challenge of quantum mechanics. 'A whole new theory of a most fundamental kind, affecting our whole outlook, has gradually been developed during the past few years,' he explained, 'and to keep pace with this alone would easily absorb my whole time . . . When these ideas become absorbed in popular science they are bound to have far-reaching effects on our views on life and the universe – but I don't want to bore you by talking shop.'[16] Freda did not mind, although she had already remarked: 'I'm afraid you dwell very much on the mountaintops and I in the plains, so far as intellectual development and attainment goes.'[17] If this was coyness he did not rise to it: 'Yes, I am dwelling more or less on the mountain tops as far as physics is concerned. The trouble is that this gives me such high standards to judge myself by, that it is easy to get an inferiority complex.'[18]

Despite the shop talk – or perhaps with its help – the relationship continued to flourish. She wrote happily of playing golf, of coping with an outbreak of scarlatina among her young pupils and of political events in Ireland (she disapproved of the up-and-coming Eamon de Valera), while he reported on his eight-mile runs, the state of his new lodgings (he had moved into Allibone's old rooms at 13 Albion Row) and a visit from his younger brother, Jim. In April they saw each other again while she was on another trip to London and though they agreed afterwards that they were both shy by nature the encounter was judged a success. Soon afterwards the Ph.D. thesis was handed in and Walton began to tell Freda of his anxiety about the *viva* – 'I never faced any exam in my life in maths or science for which I was so badly prepared.'[19] His fears were no doubt underpinned by Rutherford's reputation for asking unorthodox questions of candidates with a view less to testing their knowledge than establishing their manner of approaching a problem. Laudable as this was, it was guaranteed to unsettle.

One Wednesday at the end of May, after two days' notice ('that did not leave me very long to revise'[20]), Walton duly faced Rutherford and C. T. R. Wilson – both Nobel prizewinners – for an hour-long grilling, and ten days later he learned the outcome. Two of the five candidates had failed but as of 20 June 1931, when the degree was formally conferred, he was Dr Ernest Walton. Freda, who never had any doubts, sent her congratulations even before the result was announced and presented them in person soon afterwards when the holidays began and they shared another happy rendezvous outside Anderson & McCauley's in Belfast. By that time, back in Cambridge, the 300,000-volt machine had been dismantled for the last time and the refurbishment of Lecture Room D was under way.

Berkeley's Ernest Lawrence had something in common with Rutherford and something in common with Cockcroft. Like Rutherford he had powerful scientific insight, the gift for effective experimentation and a tremendous drive for achievement, indeed Rutherford would later say that the American reminded him of himself in his younger days. Like Cockcroft, however, Lawrence routinely heaped so much work upon himself that he could rarely see an idea all the way to completion on his own. For this reason the story of the cyclotron, though it is dominated by Lawrence's personality, is also one in which other scientists have important contributions to make. The first of these supporting actors was Niels Edlefsen, a Utah Mormon who was killing time at Berkeley after finishing his Ph.D. thesis when Lawrence asked him to build the first cyclotron. The senior man provided the drawings, the support and the motivation while the junior man wrestled with the hardware, and the upshot was the apparatus demonstrated in summer 1930 to the annual gathering of the National Academy of Sciences – the occasion on which Lawrence proclaimed that 'no serious difficulties' stood in the way of achieving disintegration with such a machine. Soon after this Edlefsen moved on and Lawrence needed someone else to take the project forward. His eye fell upon a recent recruit to the lab, a big, determined California boy with a masters from Dartmouth College in New Hampshire and a curious name: Stanley Livingston. Livingston was in search of a good Ph.D. project and entertained proposals from several possible supervisors. 'I finally selected the one Lawrence had suggested because it seemed to be more exciting; he seemed to be a very confident person with a real knowledge of the problem. He gave everyone a feeling of enthusiasm and confidence.'[21]

In the autumn of 1930, while Cockcroft and Walton were abandoning their 300,000-volt machine in favour of something more powerful, Livingston embarked on a close study of the Lawrence-Edlefsen cyclotron and quickly came to a startling conclusion: he was convinced that it did not work and never had. The success supposedly demonstrated to the academy had been, in the words of one historical account, 'an illusion of faith',[22] while the confident forecast of disintegrations was merely, as Livingston himself put it, 'a sample of Lawrence's optimism'.[23] Another man might have abandoned the project on the spot, but Livingston had faith. He soldiered on, testing and probing the Edlefsen device, discarding it, building one of his own, tinkering and adjusting the minutest details until, at the turn of the year, he believed he was achieving in reality what the boasts had claimed six months earlier. His 80,000-volt particles were feeble by the standards thought necessary for disintegration but he was satisfied that they had gained at least some speed inside the machine. It seemed to work after all, and Livingston was pleased to find that Lawrence was thrilled. 'You see, we'd proven the point. He was off to the races.'[24]

Such was Lawrence's enthusiasm that before they were half-way through with this machine he got work started on another that was to be more than twice its size. The younger man toiled long hours (there was no nonsense at Berkeley about closing at 6 p.m.) and though Lawrence could not help full-time he would call in whenever he had the chance, day or night, and was always brimming with ideas and eager for progress. The pressure on Livingston was enormous and yet in the midst of it all, like Walton in England, he found that he had to submit a Ph.D. thesis at short notice. When he asked for time to do his background reading Lawrence brushed him off. 'We're making history,' he said. 'We can go back to the basic literature any time. Just write your thesis.'[25] And all the time, sure enough, the cyclotrons were pushing particles faster and faster, and in a quite revolutionary fashion. Just as Lawrence always promised, they did so without requiring a high voltage to start with – 10,000 volts of alternating current was enough. And though the full array included a transformer, an electromagnet, the inevitable vacuum pump and other supporting equipment, the core component of the machine was a squat brass container that could be held in one hand. It worked like this.

Imagine two small, shallow metal boxes, each with an open side and both connected to a supply of alternating current. The current feeds each box alternately with positive and negative charges so that when one box is positive the other is always negative, and they constantly swap these roles. Now add a supply of protons in the middle, between the two open sides. Because opposites attract in electricity the positively charged protons will rush to the box that is negatively charged at that moment, but as soon as the a.c. switches they will want to get out of that box because like charges repel, and they will be drawn towards the other box, which is now negative. In this way, with the constant alternations, you can create a rather frenetic tennis match, with protons bouncing back and forth between the boxes many times per second. The next step is to persuade the protons to move in a circle, and this is achieved by applying magnetic force, with the poles of the magnet above and below the middle of the assembly. This has the effect of pulling the moving particles sideways once they have crossed from box to box, so that instead of hopping straight back and forth they move in a curved path. And if you can get the balance of the electrical and magnetic forces just right then a particle entering a box will be pulled through a curve of ninety degrees before the current switches, at which point it will perform another ninety-degree curve on its way out. And once it gets into the other box the same process will occur. Combine four ninety-degree curves in this way and you have particles moving in a complete circle, whirled around and around like the woman in one of those dizzying Scottish dances. Unlike her, however – and this is the really ingenious part – the protons in the cyclotron continually gain speed. The magnetic and electrical forces operate in such a way that every time a particle reaches the gap between the boxes it is pushed from one box towards the other. Twice in every circle, therefore, the protons receive this helpful shove which adds a little to their speed, and so the more circles they complete the faster they go. Their course, as they do this, becomes a steadily widening spiral, and once they have been whirled around sufficiently this spiral path takes them (by now at very high speed) to an aperture at the edge of the cyclotron where they can be drawn out and used in experiments.

Robert Millikan, the director of Caltech and the leading American physicist of these years, was among many who had thought the whole scheme impossible, muttering to a neighbour after hearing a presentation: 'It's a nice idea but it just won't work.'[26] Between them, however, Lawrence and Livingston made it work, and work well, so that in July 1931, at about the time Walton and Freda Wilson had their latest tryst in Belfast, Lawrence believed that particles were leaving the cyclotron with energies of 900,000 volts. 'The proton experiment has succeeded much beyond our fondest hopes!!!' he reported jubilantly to a friend.[27] A month later there were more exclamation marks when a telegram reached Lawrence while he was paying a visit to Connecticut. He had travelled east mainly to pursue his courtship of Mary Blumer, a young biologist and the daughter of a Yale professor, and he read the message aloud to the Blumer family. 'Dr Livingston has asked me to advise you that he has obtained 1.1 million volt protons,' wrote the physics department secretary at Berkeley. 'He also suggested that I add "Whoopee!"'[28] More convinced than ever of his destiny, and perhaps hoping that the news would strengthen his suit, Lawrence promptly ushered Mary out of the room and proposed to her.

When it came to publication the Berkeley men were more cautious, on the one hand because they were worried that their method of calculating particle energies was open to challenge and on the other, perhaps, because Lawrence was chastened by the premature claims of a year earlier. What they announced in a July letter to the *Physical Review* was dramatic enough: protons at half a million volts and 'no significant difficulties' in the way of the full 1 million.[29] That was the limit of Lawrence's caution: so far as the development of the cyclotron technique was concerned he was already thinking about 20 million volts – several times the energy of natural alpha particles. No one could guess what effects might be achieved by bombarding nuclei with such weapons. From the successful third apparatus, known as the '11-inch', which had stretched the budget of the Berkeley physics department to breaking point, he wanted to jump immediately to a 27-inch cyclotron. For this he would need a truly enormous electromagnet and he had one in his sights. Rusting away in a dump in nearby Palo Alto was an 85-ton magnet built during the First World War by the Federal Telegraph Company for use in radio transmission across the Pacific. It had been made redundant by the arrival of electrical valves and the company was ready to give it away. Lawrence's problem, not unlike Cockcroft and Walton's though on a larger scale, was that he needed a new laboratory to accommodate this monster. While he threw himself into the task of making this possible Stanley Livingston continued to work away at the 11-inch, and it was while he was pushing it up towards its limits that he suddenly encountered an unexpected obstacle. As the particles moved out into the wider spirals of the machine he could see that they were progressively losing their stability. In other words they wobbled up and down and soon went astray altogether. It was a serious difficulty all right, so serious that unless it was overcome the 27-inch would be a waste of time. In the autumn of 1931, therefore, the Berkeley men had their hands full.

At the Carnegie Institution in Washington, Merle Tuve was well aware of the progress being made on the other side of the continent. He and Lawrence still corresponded and occasionally met, so at every stage Tuve felt the full force of his old friend's optimism about the cyclotron. But his pleasures were not just vicarious because he had something of his own to be excited about in 1931. No one had worked longer and harder than he had in the search for an effective particle accelerator and few had mastered the range of techniques as fully, but his ambitions had been held in check by that albatross around his neck, the Tesla machine. The institution had not hitherto been prepared to pay for another power source capable of giving him a million volts or more and with the Depression biting in the United States it did not look as if it would change its mind. Now, however, a cheap and simple alternative to the Tesla suddenly presented itself, courtesy of a young man called Robert van de Graaff. An Alabaman with a background in mechanical engineering, Van de Graaff studied physics in Europe in the 1920s, first at the Sorbonne in Paris, where he heard Marie Curie lecture, and then as a Rhodes Scholar in Oxford. He was in England in 1927 at the time when Rutherford made his appeal for high voltages at the Royal Society and it may well have been this that set him thinking some unusual thoughts on the matter. Returning to the United States in 1929 to work at Princeton, he began tinkering with an apparatus that would put these thoughts into practice. At some stage he visited the Carnegie in Washington and Tuve would later recall a dialogue that went like this:

VAN DE GRAAFF: Merle, when I was over at Oxford and talking to Townsend, I asked him why they didn't use the old-fashioned electrostatic procedures to make a good voltage – just put up a round ball and charge it up. You ought to be able to go to a million volts.

TUVE: Yeah I know, but how? Of course a million volts sounds very interesting to me. How are you going to charge it?

VAN DE GRAAFF: Well, just run a belt into it.

TUVE: Say . . .[30]

It probably was not quite so simple as that, but the underlying facts are there: Van de Graaff made contact with Tuve after his return from Europe and Tuve encouraged him to develop his ideas because he saw in them the possibility of an alternative to the Tesla. In the summer of 1931, by which time Van de Graaff was about to move from Princeton to the Massachusetts Institute of Technology (MIT), those ideas were bearing fruit. From an early benchtop model he had progressed to a prototype comprising two copper spheres on glass pillars that were capable of producing 750,000 volts. Tuve drove to Princeton, packed the machine in the back of his car – it was small enough for that – put Van de Graaff in the passenger seat and returned to Washington. There they assembled the generator and connected it to one of Tuve's experimental tubes, especially fished out of its oil tank for the purpose. It worked perfectly, the tube withstanding 600,000 volts. For Tuve, who looked so close he got a spark to his nose, the apparatus seemed heaven-sent. 'This simple device which Van de Graaff has demonstrated to be capable of such serious possibilities,' he wrote, 'will undoubtedly alter the whole course of high-potential experiments, here and elsewhere.'[31] Best of all, it had cost just $100 to construct and would run off the ordinary mains socket in the wall.

As Van de Graaff had indicated, his machine was old-fashioned in inspiration and indeed the basic theory could be traced back more than a century. He made a sphere out of a conducting material and raised it off the ground on an insulating pillar. Inside the pillar ran a continuous belt of silk or paper – some material that could carry an electrical charge. At the bottom a charge, positive or negative, was applied to the belt from a generator by 'live' brushes and at the top it was taken off and applied to the sphere. The longer the belt ran the greater the charge that would build up on the sphere, and the bigger the sphere the greater its electrical capacity, provided it was high enough off the ground. Once the charge passed the maximum level the sphere would discharge to the nearest earthed point in a spark like a bolt of lightning. Van de Graaff's working model comprised two spheres, one of which was given a positive charge and the other a negative, so that the 'potential' between them was double the maximum of either.

From a machine such as this Tuve would certainly be able to draw the high d.c. voltages he needed.

Inevitably there were difficulties. First, Van de Graaff was not in a position simply to donate his equipment, cheap as it had been to construct. After two or three weeks he took it away to demonstrate it at a scientific gathering in New York and from there it went to his new home at MIT, where plans were already being laid for a much bigger version. Then there was the problem of size. Tuve was aiming for 2 million volts to disintegrate nuclei and to achieve this he knew he would require spheres about two metres across, raised at least three metres off the ground. This was very large to begin with, but the space to house it would have to be larger still because of the danger of sparking against a wall or ceiling that was too close. The existing laboratory was nowhere near big enough – even Van de Graaff's $100 model, which had been small enough to pack in the back of Tuve's car, could not be operated to full capacity there. So, like Lawrence and like Cockcroft and Walton, Tuve found himself in need of more space. While his colleagues Hafstad and Dahl set about designing the new machine, he made his case to the Carnegie Institution management.

Tuve's team had already achieved more than anyone could have expected with their Tesla equipment. In the spring of 1931 they had managed to produce a beam of positive particles – a jumble of protons and molecules – and by the summer they had found a way of separating these components. While Tuve estimated that at the very least these were 500,000-volt protons and they might even be 900,000-volt protons, the beam was so faint and the difficulties of observation so great that they were still virtually useless for experimental purposes. Gregory Breit, who had left Washington, continued to encourage Tuve to go ahead and bombard nuclei, but with such poor weapons there seemed to be no point. Tuve pinned all his hopes on the Van de Graaff machine.

By the latter part of 1931, therefore, the teams in Cambridge, Washington and Berkeley were rushing along parallel courses and all three, at almost the same moment, had encountered the same banal obstacle. In Cambridge they needed Lecture Room D to house their four-metre-high tower of rectifiers; in Washington they needed a home for the big Van de Graaff spheres; at Berkeley accommodation had to be found for an 85-ton magnet. This was no coincidence, for as these scientists were probably aware they were pioneers in a change of culture. Thirty years earlier the curtain had finally come down on the age of the gentleman scientist and the task of fundamental discovery passed to universities, research institutes and commercial laboratories. But while the location and funding of the work changed, its essential character did not, so that by and large researchers continued to work alone or in pairs and to use small-scale, handmade apparatus that could fit on a workbench. So it had been with Rutherford and the Curies, and so it still was with Chadwick, Geiger, Meitner and Bothe. The construction of particle accelerators was among the first signs that this was about to change; big physics was on its way.

12. Timeliness and Promise

When Cockcroft and Walton moved into their new room in the Balfour Library building in the first days of August they were in a hurry, although at that moment the imperative was less scientific than social. In two months' time the Cavendish was due to celebrate the centenary of the birth of its founding director, James Clerk Maxwell, and Rutherford, whose sense of history was strong, was determined that no effort would be spared in making the event memorable. Cockcroft (who else?) had been given the task of organization and he had invited some of the greatest names in physics: Planck from Berlin, Bohr from Copenhagen, Millikan from California, Langevin from Paris, Guglielmo Marconi from Rome, Pieter Zeeman from Amsterdam and others. There would be lectures, ceremonies and dinners in both London and Cambridge and of course the guests would tour Maxwell's laboratory and see the work being carried on by his successors. When it came to showing off of this kind, Rutherford had long before realized that the large, rumbling, sparking Cockcroft–Walton apparatus was among his most impressive assets and the two scientists had often laid on demonstrations in the past. It was vital, therefore, that they should have something to show their distinguished visitors in October, and since the lab would be closed in September because of holidays this had to be achieved in just a few weeks.

Most of the construction work fell once again to Walton because after barely a week Cockcroft left on a trip to the Soviet Union. It was a working visit with a mixed party of scientists and others, organized by the *Manchester Guardian*'s science correspondent, J. G. Crowther. (This was the same Crowther who had commissioned Gamow's book on the nucleus; we shall meet him again.) Cockcroft was not left-wing but he was curious about revolutionary Russia and he no doubt received encouragement from his friend Kapitza to go and see what was being achieved under communism. Metropolitan-Vickers was also eager to hear his views; they had long-term business links with Russia and a permanent office in Leningrad, and they subsidized the trip.

He sailed from London on 13 August and his tour took in Leningrad, Moscow, Kiev and Kharkov. This was the time of Stalin's first Five-Year Plan, a period of activity and excitement that appeared in sharp contrast to the worrying economic stagnation of the West, and all of the best that the Soviet system could offer was on display. Cockcroft saw universities, scientific institutes, power stations, dams and factories and had the opportunity to talk to everybody from construction workers to senior politicians. He came away impressed with the youth and energy of those he met but still doubtful about the politics that lay behind it all. 'Whether the final result will bear any resemblance to the Marxian communism is one of the most interesting questions for the future,' he concluded.[1] The trip also had its social side as he was able to call in on the Metro-Vick office, pay a visit to Kapitza's mother and meet at least one recent colleague from the Cavendish, Yuli Khariton, now working in Joffe's laboratory in Leningrad. George Gamow was also in the Soviet Union at this time but was in Odessa that August visiting his father and so did not cross Cockcroft's path. This was probably a pity, for the Russian might have passed on his suspicion that the climate of Soviet science was becoming more restrictive, a suspicion confirmed only weeks later when Gamow was refused permission to attend a conference in Rome. Just as the Englishman visited, in fact, the period of Soviet scientific openness was coming to an end.

When Cockcroft returned to Cambridge in early September Walton was back in Ireland again, but he found that much had been achieved and was able to push the work forward alone. The new rectifier column, an imposing pillar of glass comprising four cylinders each a metre in length and all manufactured to order by their suppliers in Jena, stood complete in the middle of the room, with the old transformer, generously moved and reinstalled by Metro-Vick technicians, against a wall to one side. At least some of the condensers were in place, and also the shiny metal 'spark-gap spheres' that were among the simplest and yet most spectacular parts of the apparatus. There were two of them, each nearly a metre across, one resting on an insulating pillar and the other strung above it on adjustable ropes from the ceiling. When the voltage multiplier was charged and the upper sphere was lowered, at a certain point a very loud and brilliant spark would jump the gap. It was a tool to allow Cockcroft and Walton to measure the voltage but it was also a party trick capable of pleasing even seasoned scientists. It duly did so as, in the first days of October, the laboratory paid tribute to its founder. The foreign dignitaries were joined by the cream of British physics, including Sir Oliver Lodge, Sir James Jeans, Sir Ambrose Fleming and from Cambridge Sir J. J. Thomson and Lord Rutherford, Clerk Maxwell's surviving successors. 'Never before have so many famous scientists circulated in the Victorian rabbit warren of the Cavendish,' noted one who was there.[2]

It is unlikely that the acceleration tube and the proton source were on show for the visiting dignitaries, and the fireworks displays were probably restricted to relatively modest voltages. Once the festivities were over, however, the work resumed and the whole machine came together with a speed that was remarkable given that no full-scale testing had been possible before the move – testimony, no doubt, to an increasing theoretical mastery of the problems involved. A welcome boost at this stage was the performance of the plasticine. Now they were using the big glass cylinders in earnest, connecting and disconnecting them as they sought the best arrangements for the internal parts, they were delighted to find that this durable putty beloved of small children more than matched the performance of sealing wax as a scientific tool. 'The ease with which the joint may be made and broken again make it a very convenient type of vacuum joint,' they would write.[3] For example, a flat plate could 'easily' be joined to a brass tube 'even if the end of the latter has merely been sawn off roughly with a hacksaw'.[4]

As the tests progressed and the stresses grew greater, however, this confidence was punctured, for now the plasticine was sometimes sucked into the tubes by the high vacuum. Cockcroft had the idea that a better quality putty might help and he suggested to Metro-Vick that their new Apiezon oil, which had so transformed the business of vacuum pumping, would do very well. He was right: incorporated into a paste, the oil proved itself effective beyond their dreams. They were able to drill holes in the glass tubes to feed in wires, seal them with the new mixture – Apiezon Compound Q, it was called – and still it was possible to hold a good vacuum. Of course there were always leaks of some sort and both men spent a lot of their time atop ladders with tubes of paste, pushing it into the gaps with their thumbs and then coating the exterior with grease. But they knew how much worse it could have been, for the whole arrangement would have been a near-impossibility if they had had to rely on Bank of England sealing wax. Using Compound Q, if an internal fault developed in the glass tower the whole thing could be dismantled and reassembled in a couple of days.

We have glimpses of the progress in this autumn term from Walton's letters to Freda, which continued to extend in length and broaden in variety. Of the Maxwell centenary he wrote with pride of how many of the great men turned up to see him demonstrate 'my apparatus'. In mid-November he reported 'a rather bad day of it in the lab. A heated filament inside my apparatus broke and in order to repair it, the whole thing had to be dismantled. I got it together again today and am now not quite as far on as I was two days ago. However, we are used to that sort of thing in the Cavendish and it is often described as negative progress.'[5] When Freda observed that such things must tax his patience he replied that he was not the sort to 'blow up' when things went wrong – 'I am not an explosive or temperamental person'.[6] And in the same letter he was able to say that 'things have been going very well, or at least as well as can be expected'. Inevitably his scientific partner entered the correspondence: 'Cockcroft, with whom I work, is an appalling writer and I usually have to use all my ingenuity to make head or tail of what he writes.'[7] (This was a difficulty he shared with everybody but Cockcroft himself.) A little later there was a mention of the feverish building work going on around the Cavendish. 'Cockcroft has had a lot to do with the plans etc. for this, with the result that I have had to do nearly all the work in our new room by myself for the last few months.' This was not a complaint, however, and it was followed later in the same letter, dated 8 December, by the announcement: 'The apparatus that I have been making since the summer was working today for the first time and behaved almost exactly according to calculations so I am feeling quite pleased with life. Of course something may go wrong at any time. I hope not.'

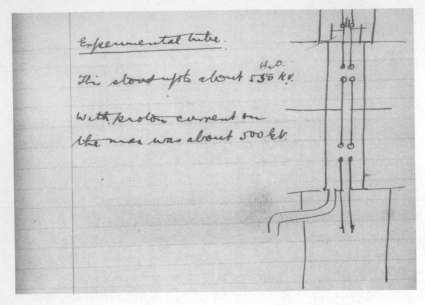

Experimental tube.

This stood up to about 550 k.v.

With proton current on
the max was about 500 k.v.

A Walton sketch for the acceleration tube, with proton
source above and observation hut below

Here was a milestone. More than three years had passed
since they started work and now, in the big, bright new room,
all the pieces were finally in place. The transformer, condensers and rectifiers of the voltage multiplier were ready to
be tested to something approaching their limits. The Metro-
Vick pump for the rectifiers had been installed under the
floor. The separate transformer for the proton source resided
atop its white porcelain insulating column, wrapped in a
protective metal shield and connected by its loop of rope to
the motor on the floor. A little apart stood the acceleration
tube, unrecognizable now from the arrangement used in the
basement room. To handle the higher voltage it comprised
two metre-long tubes of the cylinder type, one above the
other, and two sets of electrodes inside them, still in the form

of pipes. At the top, housed in what appeared to be a huge tin top hat, was the proton source, and at the bottom was the experimental chamber in a wooden hut now sufficiently commodious to seat one person in reasonable comfort. The lead covering was more ample and there was a little curtain of black cloth to improve viewing conditions inside. To the side was another Metro-Vick pump to evacuate the acceleration tube, while in the corner of the room, by the door, was a desk streaming with wires and decked with switches and gauges, which was impressively referred to as the control table. With the spark gap spheres by the window and the great columns of the rectifiers and condensers jutting towards the ceiling, the room had an appearance somewhere between a metallic forest and a set from *Metropolis*. The next step was to test the machine, but that process was soon interrupted by the Christmas break. And when the Cavendish staff reassembled in the new year of 1932, the year that would go down in history as the *annus mirabilis* of nuclear science, it was James Chadwick and not Cockcroft and Walton who stepped into the limelight.

As we have seen, while Rutherford oversaw the Kapitza work and the accelerator project at the Cavendish, Chadwick had remained closer to the continuing research using radioactive particle sources. In particular he had nurtured the efforts to harness polonium together with the new counters in a fresh attack on the nucleus. One morning in January 1932 he picked up the scientific journal *Comptes Rendus de l'Académie de Science*, just arrived from Paris, and came across an article by Frédéric Joliot and Irène Curie. The French couple had been as impressed as he was by the findings of the German researchers Bothe and Becker a year or so earlier and set out to replicate the experiments in just the way that Chadwick had asked young Hugh Webster to do. Unlike the Cavendish, however, the Radium Institute in Paris possessed the world's largest stock of polonium and had long experience in handling it. Joliot and Curie, moreover, applied a good deal of ingenuity to the matter and the result of their investigations was much more dramatic than anything Webster had been able to find.

Bothe and Becker had shown that when they aimed the alpha particles from their polonium at targets of the light element beryllium some of the particles entered and merged with beryllium nuclei, and then these nuclei in their turn emitted very powerful rays. Their interpretation was that these were gamma rays carrying off a surplus of energy. When Joliot and Curie looked into this they were able to replicate the effect without difficulty and so observe the rays, and they went on to see what those rays could do. One trick they tried was to place in their path a sliver of paraffin wax, a substance composed largely of hydrogen atoms, and this produced another intriguing result, for when they examined what was happening on the far side of the wax they found a shower of protons. What could this mean? So powerful were the new rays, it seemed, that they were capable of knocking protons out of paraffin wax. And not only did the protons reel away, but they did so at high speed. The idea that gamma rays could behave in such a way was not entirely unfamiliar: just like light, gamma rays were known to move in quanta (eggs rather than milk) which were known as photons, and these were capable of behaving as particles. Photons were known on occasion to shift electrons by colliding with them. Here, however, it appeared that they were actually shifting *protons*, which were almost 2,000 times more massive than electrons. Applying well-known formulae, Joliot and Curie sat down to work out just how much energy gamma-ray photons would need if they were to affect protons in this way. The correct unit for measuring energy in this form was the 'electron volt' and as a yardstick natural alpha particles escaping from radium had energies above 7 million electron volts, while Cockcroft and Walton were aiming to push their protons to around 800,000 electron volts. When the French couple completed their calculations they found that the rays they had observed must have energies some way off this scale, at around 55 million electron volts.

Reading this in his first-floor office Chadwick found the conclusions 'most startling'[8] and he was soon joined by an equally astonished Norman Feather, fresh from reading the same paper in the library. 'Just nonsense, isn't it?' said the younger man. 'Yes, just nonsense,' came the reply.[9] The incredulity was understandable because no atomic phenomenon of this kind – nothing remotely like it – had ever been detected before. It was as though a football kicked against a lorry had been able to impart enough energy to propel the lorry for several miles. But while Feather was simply dismissing the French claims Chadwick was thinking more deeply. When he spoke to Rutherford later that morning there was the same reaction – 'I don't believe it' – but there was also a shared view that this was not just a laboratory cock-up; Joliot and Curie were too good for that. 'Rutherford agreed that one must believe the observations,' recalled Chadwick later. 'The explanation was quite another matter.'[10] Given their shared history both men must have had the same suspicion but it was Chadwick who went off and put it to the test.

What followed was a series of experiments still admired as models of ingenuity and rigour and then a published paper that could scarcely have been more conclusive. And the essential work took just three weeks. The French paper, as it happened, could not have landed at a better moment for Chadwick. As part of the same reshuffle that had taken

Cockcroft and Walton to Lecture Room D, he too had moved
house within the Cavendish over the course of the winter and
his research had been interrupted. When he settled into his
new room his first act had been the laborious one of extracting
polonium from the tubes given to Feather by the Kelly Hospital in Baltimore. This 'very, very fine source',[11] as he called
the finished product, was the first truly satisfactory quantity of
polonium ever prepared at the Cavendish and it was just ready
that January. Had the Joliot-Curie paper been published a
couple of months earlier he would not have been in a position
to react but as it was, equipped with his new source and some
of the latest counting apparatus of the Wynn-Williams type,
Chadwick now threw himself into the most determined and
confident of all his campaigns to pin down the neutron.

Powerful as it was, the polonium source was no more than
an all-but-invisible film of matter spread over the surface of
a silver disc the size of a modern penny. This Chadwick
placed close to another, slightly bigger disc of pure beryllium
and then enclosed both in a small container emptied of air.
Inside this the alpha particles emitted by the polonium atoms
rained down upon the beryllium, penetrating its nuclei and
causing them in turn to emit their peculiar rays. These rays
Chadwick allowed to stream out of one side of the container
so that in effect he had created a portable gun, continuously
firing the rays. He first aimed this gunfire into a cylinder of
gas attached to a counter, recording the results using the
wobbling-mirror method with the moving photographic
paper. There was a small effect, with a few 'kicks' per minute
showing on the paper. Next Chadwick followed the example
of Joliot and Curie and inserted a two-millimetre sheet of
paraffin wax between gun and counter. Suddenly the kicks
were far more frequent: this was the effect described by the
French couple. While most of the rays were passing clear

through the wax (which, like all atomic material, was largely empty space) some were crashing into the hydrogen nuclei (protons) and knocking them clean out. The size of the kicks left no doubt that they were protons, just as the French couple had said.

Having confirmed that the rays were indeed powerful enough to knock protons out of wax, Chadwick now measured the energy of the displaced protons. By placing sheets of aluminium foil of varying thicknesses in their path he established the minimum thickness capable of stopping them, and from this he was able to calculate that they left the wax with an energy of 5.7 million electron volts. Returning to the analogy of the football kicked against the lorry, Chadwick knew from this that there was indeed a lorry and that something had struck it which caused it to move, say, ten miles. What he did not know yet was whether that something was really a football, in other words whether it was a gamma ray. So he looked at some other effects caused by the rays. He aimed his gun at some of the light elements, notably lithium, boron, carbon and beryllium itself, to see what would happen. In every case the counter recorded activity which Chadwick interpreted as the recoil of atoms of these materials when struck by the rays. They did not just dislodge a proton; instead they gave the whole atom a shove. Chadwick tried aiming the rays into gases – hydrogen, helium, nitrogen, oxygen and argon – and again he detected those recoil atoms. 'It appears then,' he wrote, 'that the beryllium radiation can impart energy to the atoms of matter through which it passes and that the chance of an energy transfer does not vary widely from one element to another.'[12] In other words, the rays delivered a big punch no matter what they bumped into.

This business of putting the rays through their paces against ten or twelve different elements is easily described but was not easily achieved. The 'gun' at the heart of the apparatus was simple enough, a small metal cylinder with an evacuation pipe, clamped crudely to a piece of recycled timber (still preserved in the Cavendish, it resembles nothing so much as a piece of discarded domestic plumbing). This was linked to secondary vessels in which the various ray effects took place, and they had to be meticulously cleaned, treated, sealed and re-calibrated with every change of target material. And then there was the counting equipment, still cumbersome for all the work that had been done on it. Long series of readings were conducted on each target, with frequent repetition if anomalous results appeared, all requiring the development of the endless film strips and the patient counting of kicks. Chadwick had his own lab assistant, Horace Nutt, who carried his share of the burden, but the physicist was now working flat out – even to the point of reopening the laboratory after dinner so that he could carry on into the night. In the heat of the chase, it seems, the frailty which had haunted him since the war was forgotten and his attitude is summed up in a much quoted exchange from those days: 'Tired, Chadwick?' asked a colleague. 'Not too tired to work,' came the reply.[13]

Across the range of elements he calculated in each case the energy imparted by the rays to the recoil atoms and he found, for example, that the atoms of nitrogen moved away with an energy of 1.2 million electron volts. Joliot and Curie had established that, to produce the effect on protons which they had witnessed, gamma rays would need an energy of 55 million electron volts. But Chadwick's nitrogen collisions were even more remarkable because nitrogen atoms are fourteen times heavier than protons. This football, it seemed, was capable not just of moving a lorry, but of shifting a whole house. In fact Chadwick calculated that to cause a nitrogen atom to recoil in the way he had observed, a gamma-ray photon would need an energy of 90 million electron volts – a mind-boggling figure. And on the evidence he was able to gather it appeared likely that even more extraordinary totals would emerge from experiments with heavier elements.

Here at last Chadwick drew the line. It was too much for a reasonable scientist to accept, indeed it would be in flagrant breach of several basic laws of physics, to suggest that rays of such power could exist in and escape from the nuclei of beryllium or any other element. A choice presented itself: either you must suspend the application of the laws of physics or you had to think about the rays in some other way. Chadwick knew which he preferred, as he would write soon afterwards: 'If we suppose that the radiation is not a quantum radiation [that is, not photons of gamma radiation], but consists of particles of mass very nearly equal to that of the proton, all the difficulties connected with the collisions disappear.'[14] In other words, this was not a football hitting a lorry but something similar to a lorry hitting a lorry, and in those terms the consequences suddenly made sense. And more: 'In order to explain the great penetrating power of the radiation we must further assume that the particle has no charge.'[15] That would explain why it did not seem to be troubled by the usual repulsive forces around the nucleus. What were these particles, similar in weight to protons but with no electrical charge? They were of course neutrons, those solid, deadweight entities whose existence Rutherford had predicted twelve years earlier. They had real substance, just like protons, and because they had no positive charge (as protons and alpha particles do) they would not be repelled or deflected by nuclei and so could penetrate them with relative ease – hence the punch they packed. Going back over the various experiments Chadwick could show that everything fitted, that with neutrons in the equation every single collision obeyed the fundamental laws. The polonium–beryllium effect at the heart of it was revealed as this: the polonium emitted an alpha particle of weight 4 which struck a beryllium nucleus of weight 9, and what emerged from the collision was one whole nucleus of weight 12 – which is carbon – plus a single, spare neutron of weight 1. Thus the mysterious rays observed leaving beryllium by Bothe and Becker were ejected neutrons (as well as some incidental gamma rays) and it was these neutrons that had knocked the protons out of Joliot and Curie's paraffin wax.

At last, Chadwick had discovered the neutron. All through this series of experiments he had given the benefit of the doubt to the gamma rays – they were after all known phenomena, while the neutron was no more than an idea. But as the results stacked up he showed in his cautious way that the only entity capable of causing them was something that weighed about the same as a proton and had no charge: by definition the neutron. Having finally pinned down his prey and stripped it of its infuriating camouflage, Chadwick was determined to weigh and measure it, describe it from every angle and probe it as deeply as all the techniques of the Cavendish would allow. The work went on for months but he did not wait to announce his discovery. On 17 February he sent off a short note to *Nature* entitled 'Possible Existence of a Neutron' and a few nights later, while the journal was still at the presses, he described his experiments and conclusions to an audience of his colleagues and students. Warmed by success and a good dinner at Trinity College, Chadwick was for once in relaxed mood as he described the various stages of what one present called his 'long quest',[16] and the climax of the story was greeted with an ovation both spontaneous and excited. Resuming his seat the speaker declared: 'Now I want to be chloroformed and put to bed for a fortnight.'[17] Rutherford was delighted in every way. He was pleased for his colleague, proud of his laboratory and more than gratified that his own prediction had been vindicated. Above all he was thrilled to see a veil so decisively removed from around the nucleus. It was truly *bahnbrechend* stuff. What was happening, meanwhile, with his own pet project?

The tests that Cockcroft and Walton carried out on their machine before Christmas and in the first days of the new year were encouraging. It took a full day to warm up the voltage-multiplying components because once the rectifier tower was cleared of air and the first voltages were applied they found that various gases 'evolved' from the glass walls inside. Cloudy, coloured glows of green and blue would appear. The voltage had to be held steady while these were pumped off and then it could be raised gently until further gas traces appeared and these too were removed. Once this step-by-step ritual of 'outgassing' was complete they were pleased to find that – at least until it next had to be dismantled – the apparatus maintained its integrity very well. The vacuum was sound and leakage was minimal even when everything was switched off overnight. In the mornings, therefore, all they had to do was turn everything on and wait for the pumps to restore the full vacuum and they could proceed to full voltage within half an hour. This full voltage was not yet at the level they were aiming for because one condenser was not up to its job and until it was replaced their ceiling had to be 700,000 volts. Even then, they could see that the acceleration tube was under extreme strain. The first time they applied 700,000 volts the old problem of surface creepage returned, a spark appeared on the glass and the lower of the two big cylinders was punctured. Thanks to the new Compound Q mixture they were able to plug the hole and resume work in short order but every time they went above 700,000 volts it happened again. For higher voltages they realized they would need longer glass cylinders and these they promptly ordered from Germany, but in the meantime they had to settle for an upper limit of rather less than the desired 800,000 volts.

As these tests proceeded a series of notices in the *Physical Review* told of further developments in the United States. First it was Robert van de Graaff giving an account of his 1.5-million-volt generator and observing, in case there was any doubt: 'The application of extremely high potentials to discharge tubes affords a powerful means for the investigation of the atomic nucleus and other fundamental problems.'[18] This was soon followed by the latest word from Tuve's lab in Washington, serving notice that 'we hope in the near future to undertake a program of quantitative measurements using the newly developed Van de Graaff generator'.[19] And next came Lawrence and Livingston, with the characteristically dramatic declaration that they could now be certain their cyclotron had produced 1.1-million-volt protons. They had solved the problem of particle instability, or wobble, and were confident that their technique 'has now been brought to a stage of development where it can serve in experimental studies of atomic nuclei'.[20] As if that was not enough, they added: 'A larger apparatus is under construction.'

Published papers were not the only source of intelligence available to the Cavendish scientists. American visitors occasionally brought reports, and some of the scientists corresponded with friends there. One such was Cockcroft, who in January 1932 received a long letter from Joseph Boyce, an American he had befriended when they were neighbours on the Cavendish Nursery course eight years earlier. 'I have just been on a very brief visit in California and thought you might be interested in a brief report on high voltage work there and in the eastern U.S. as well,' wrote Boyce.[21] His first port of call had been Pasadena, where Lauritsen was continuing work with his powerful X-ray apparatus and waiting for General Electric to supply him with a new transformer that would enable him to go to higher voltages. 'But the place on the coast where things are really going on is Berkeley,' declared Boyce with some excitement. 'Lawrence is just moving into an old wooden building back of the physics building where he hopes to have six different high-speed particle outfits. One is to move over the present device by which he whirls protons in a magnetic field and in a very high frequency tuned electric field and so is able to give them velocities a little in excess of a million volts.' This was the 11-inch cyclotron. Boyce also described several other acceleration schemes being developed at Berkeley, including a linear one along the lines Walton had experimented with in late 1928, and then he mentioned Lawrence's plan for a 45-inch cyclotron 'with which he hopes for at least 3 million volts'. The letter continued: 'On paper this sounds like a wild damn fool program, but Lawrence is a very able director, has many graduate students, adequate financial backing, and his work so far . . . has achieved sufficient success to justify great confidence in his future . . .' Nor had Boyce finished there. 'Back in the east,' he reported, 'Tuve at Washington is working on the development of tubes to stand high voltages and has ordered a six-foot sphere to build a one-ball Van de Graaff outfit for about 3 million [volts]. I think I sent you clippings about Van's own results and plans . . .'

The Americans, it was clear, were fairly galloping along. For their part Cockcroft and Walton were sufficiently pleased with progress in January to join the flurry of publication, so while their days were spent working on the proton beam their evenings were given over to writing. First they prepared a short note for *Nature* which appeared on 13 February and served as a preliminary response to what was going on in the United States. In four paragraphs they described their apparatus in outline, gave their first findings on the properties of the proton beam and declared, in terms reminiscent of Lawrence's similar notes, that 'we do not anticipate any difficulty' in running the machine up to 800,000 volts.[22] As soon as that was sent off they set about writing a longer account for the *Proceedings of the Royal Society*. Walton was again burning the midnight oil and his report of this to Freda provoked a teasing rebuke about his 'disgraceful hours' and another little passage of fond admiration: 'So you are writing a paper for the Royal Society. Will you be delivering it in person in London? You know when I think of you with your Royal Society papers, inventive ability etc etc, I feel very glad to represent even a little corner of your interests . . .'[23] Again he would have none of it: 'Please do not talk of me as having inventive ability etc etc . . . I have come across so many people who are so much superior to me in every respect that it makes me feel awkward when anyone starts talking of me in that stream.'[24]

Cockcroft and Walton were also caught up in another time-consuming project, a distraction for which Walton was largely responsible. The previous autumn, with Metro-Vick's support, Cockcroft had taken out a patent on the voltage-multiplying circuit on the grounds that it might have commercial possibilities. Since then Walton had seen that the system as they had developed it in the laboratory might have important uses in the power industry and, with encouragement from his supervisors at the Department of Science and Industrial Research (DSIR), he decided they should seek a second patent. The application was a fiddly business involving several drafts and the mediation of both Metro-Vick and a London patent agent, and the correspondence dragged on through

February and March. (In fact it was all a waste of time, in part because unknown to Cockcroft his original work was not new – the Swiss scientist Greinacher, whom we met in connection with the concept of the valve counter, had designed a similar multiplying circuit in 1921. As a result neither patent was honoured overseas and even in Britain the technology was never used commercially.)

It may have been at Rutherford's prompting that they tried their first proper experiments with the new apparatus. It was late February, with the excitement about Chadwick's discovery at its peak, and their first thought was to attempt something with beryllium, the element that had been at the heart of that story. Where Chadwick and others had bombarded beryllium with alpha particles they decided to see what effect their protons would have, so they placed targets of beryllium filings at the foot of the acceleration tube and pushed the machine up in steps to 550,000 volts, which at this stage was their working limit. They detected nothing, and when they tried a solid lump of beryllium the result was no different. Over a couple of weeks they attempted a variety of other experiments, looking for gamma rays from both beryllium and lithium and also testing the beam on helium gas, but no findings of interest turned up. It was disappointing. Easter was late that year and the lab remained open well into March, so they took the opportunity before the holiday to rebuild the acceleration tube with the new, longer cylinders, a fiddly task as ever, involving a great deal of work making joints and plugging leaks.

Besides all their other concerns in these weeks, both men were once again dealing with important developments in their personal lives. Cockcroft learned that Elizabeth was pregnant, news that, after so much heartbreak, must have prompted almost as much anxiety as it did pleasure both at home and among family in Todmorden. The baby was not due until the autumn and there was not much he could do for the moment besides take good care of his wife and think about getting the small bedroom in order. Walton, meanwhile, was experiencing a career crisis. Since he had switched from the 1851 Commission scholarship to the DSIR award he had been funded from the public purse and this now exposed him directly to the effects of the Depression. In autumn 1931 the scale of the economic crisis was finally acknowledged in Britain when the government announced a package of cuts in public expenditure. For some time it looked likely that Walton's annual grant would be reduced by 10 per cent, which would have made life difficult, but in November he learned to his relief that he would be spared that fate. Instead he was told that his second yearly increment, worth £25 and due from the following summer, was to be cancelled. This was bearable but the scare unsettled him – who was to say that this round of cuts would be the last? – and with his Ph.D. complete and a lengthy paper on the new accelerator on its way to the Royal Society he began to think seriously about his next job.

A prospect presented itself in March when an old friend from Trinity College, Dublin sounded him out about a post in the physics department there. Walton broached this with Rutherford, who did not like the idea. 'He was not at all satisfied with the status I would have,' Walton informed Freda. 'He said that if I would like to stay in Cambridge another year he would make my income up to £400.'[25] This offer, a large jump from the £300 he was receiving from the DSIR, soon improved to £450 per annum over three years but it seems that Walton was not quite convinced; though there is no hint of it in his letters to Freda he may have been tempted by the thought that long-term, salaried employment in Dublin might allow him to marry. The Provost of TCD happened to visit Cambridge at this time and was invited to Sunday tea at the Rutherfords' with Walton present, but nothing was decided and when Walton – at his professor's prompting and with Freda's support – concluded that he should at least 'hold out for a Fellowship'[26] at his old university, the word swiftly came back that he would be allowed to sit the Fellowship examination the following year.

Here was one of those moments, pleasing and disconcerting at once, when a person discovers that his market value is higher than he thought. Walton's modesty about his own talents and the experience of seeing his increment cancelled were not likely to engender confidence, but now he watched as two universities fought over him. He was not sure what he should do, though he confided in Freda that he thought 1932 would be his last full year at Cambridge. Rutherford's determination, however, was not to be underestimated. The minutes of the DSIR's scientific grants committee for 13 April 1932 show what his next move was:

The Committee considered a statement on the position of Dr E. T. S. Walton, the holder of a Senior Research Award which was due to terminate on 30th September 1933. The Committee agreed that it was important that Dr Walton should be enabled to remain at Cambridge for two more years and that the circumstances were so exceptional that after 30th September 1932 Dr Walton should be treated on the basis of a man engaged on work of special timeliness and promise rather than the holder of a Senior Research Award, and further agreed that for two years from the above date he should receive a payment at the rate of £250 per annum, to be added to the Clerk Maxwell Studentship of £200 which it was understood would be awarded to him.[27]

This is a forceful testament to Rutherford's resolve. He did not ask the DSIR for more money – at a time when other grants were being cut that was not possible – but on the basis of the 'exceptional' circumstances and the 'special timeliness and promise' of Walton's work he secured him a two-year extension of his grant. And he had decided to top this up with the Clerk Maxwell award, which was effectively in his gift and was viewed in the laboratory as a mark of special favour (Chadwick was among the previous holders). By the time Walton came to consider this, however, events had moved on.

At the end of March he travelled to Coleraine to spend Easter with his family. It was a short break but there was just time on the return trip to see Freda in Belfast before he took the night boat back to England. He broke his journey in Manchester, paying a call on Allibone at the Metro-Vick lab, and reached Cambridge on the evening of Thursday 7 April. The following day he had an appointment at the National Physical Laboratory in Teddington, west of London, where he had arranged to talk to the chief metallurgist, so it was Saturday morning before he was back in the lab in company with the big machine. After more than a week out of action it was doubtless in need of some care and attention, while there was also another paper in the *Physical Review* to be read. Lawrence and Livingston gave their fullest account yet of the cyclotron work, announced that they had achieved 1.22 million volts and spoke of reaching an astonishing 10 million volts with the bigger machine they were constructing. They also quoted Gamow, showing they were aware of the implications of his work, and they declared grandly that particle accelerators were 'probably the key to a new world of phenomena, the world of the nucleus'.[28] When they would be ready to turn that key they did not say.

13. Red Letter Day

After a quiet Sunday Cockcroft and Walton probably spent Monday outgassing the machine again in readiness for further tests and on the Tuesday they resumed the examination of their proton beam, applying a magnetic field and measuring the deflections it caused. It must have been on the Wednesday that these investigations were interrupted. Various accounts have been given of what happened and they do not agree in every detail, but the fullest is by Vivian Bowden, a young researcher then working in a neighbouring room:

Rutherford came into the lab one day. First of all he hung a wet coat on a live terminal and gave himself an electric shock, which didn't improve his temper. Then he sat down and lit his pipe – he always smoked very dry tobacco, so when he lit it, it went off like a volcano with a great big cloud of smoke, flames and piles of ash. Then he summoned Cockcroft and Walton and asked them what they were doing. He told them to stop messing about and wasting their time and go on and do what he'd told them to do months ago, and arrange that these protons were put to good use.[1]

These words were written half a century later and for that and other reasons should not be taken too literally – it is not clear, for example, how or why Bowden would have witnessed this scene. But with or without the electric shock and the smoke the general thrust – that an irritated Rutherford told the two men to get on with it – is correct. Chadwick, in a letter written some years before Bowden's account, provides some background:

One day JDC [Cockcroft] told me that all was well and they had got a beam of protons. A morning or so later I went over to see them, being naturally deeply interested. I found that they had got a nice beam but, to my horror, they were not using it for the purpose for which all their work was a preparation . . . I asked why they were not using the beam to bombard a light element; their reply was that they wanted to justify their work by getting some immediate definite results, or something to that effect. I said, perhaps rather angrily, that they were not getting on with their real job and that I should go at once to tell Rutherford. This I did, and of course Rutherford was incensed and, as soon as he could, he went to tell them to get on with bombardment experiments.[2]

Both recollections leave the impression that the two researchers had been needlessly dallying over their proton beam and it is only fair to correct this. The laboratory had been closed over Easter and before that they had rebuilt the entire acceleration tube to a new specification. Walton had also spent several days in bed with bronchitis while Cockcroft, as ever, had many commitments elsewhere. It was at least three weeks, therefore, since they had had an opportunity to experiment. Nor would it be true to say that they had neglected to bombard elements since they had done so both in February and March, without success. As Cockcroft later wrote, it was precisely because of those setbacks that they 'set to work in some discouragement to measure our proton ranges'.[3] But Rutherford's humours did not answer to reason or take account of the discouragements of others and that he was impatient on this occasion is beyond doubt. Cockcroft remembered that he 'came in and grumbled about our wasting time'[4] and Walton that he 'told us we ought to put in a fluorescent screen and get on with the job, that no one was interested in exact range measurements'.[5] There was no gainsaying the professor when he was in a mood, so they immediately began rearranging things at the foot of the acceleration tube, removing the range measurement equipment and installing a target and a zinc sulphide screen. For their first target it was natural to choose lithium, the soft, silver-white alkali metal that is the lightest of all the solid elements. The sample was placed on a small platform inside the experimental chamber at the base of the acceleration tube, directly in the path of the beam, and the platform was tilted at an angle of forty-five degrees to the vertical. At the side of the chamber, facing this tilted target, was an aperture covered with a foil that would screen out any scattered protons from the beam but still allow through any bigger or more energetic particles. Behind the foil was the scintillation screen, the familiar plate of glass coated with zinc sulphide. And behind that in turn was the microscope, ready to observe any activity on the screen. These components had to be prepared, installed, sealed and outgassed, and the job took the rest of the day.

The next morning, Thursday 14 April, Cockcroft had commitments in Kapitza's laboratory so Walton was on his own. He warmed up the machine in the usual way, first turning on the air pumps, the motors and the transformers and then, amid the general rumbling and hissing, pushing up the voltage step by step. The coloured glow inside the rectifiers told him when gas was present and he waited while the pumps removed it before taking the machine higher. After half an hour or so all was well and the voltage had reached a level Walton would later describe as 'reasonably high'.[6] Next he turned on the proton supply and particles began to flood into the top of the tube and race in a narrow beam away from the anode down through the cathode and on to the lithium target. Once he was satisfied everything was working smoothly he decided to see whether anything of interest was happening. Leaving the control desk in the corner of the room he crawled on all fours across the wooden floor – a precaution to avoid electrocution – and crept into the observation hut. Sitting on the little stool and putting his eye to the microscope, he was amazed to see the scintillation screen virtually aglow. It was not a matter of one or two little flashes per second but dozens, maybe hundreds. There were so many, in fact, that they could not possibly be counted individually.

It is an oddity of Walton's career that, perhaps alone among the younger scientists at the Cavendish, he had never before seen flashes of this kind. It was probably because of his late arrival that first term back in 1927 that he missed the routine training exercises in the Nursery, and the need had never arisen since then. For this reason he could not relate what he was now looking at to any previous experience of his own, but from what he had read and heard he was all but certain that the flashes must denote the presence of alpha particles. After watching for a few moments in mounting excitement he realized that he must do some simple tests, so he eased his way out of the box, crawled back across the floor to the control table and switched off the power to the proton source. This left the machine as a whole still running, but with no protons entering the tube. Once again on his knees, Walton returned to the microscope. Nothing. He turned the proton stream on again: intense flashes of light. Off again: nothing. It was scarcely credible.

One of the recent changes at the Cavendish that accompanied all the construction work was the installation of an internal telephone system. Up to then all communication within the sprawl of buildings had entailed journeys along corridors, up and down flights of stairs and across courtyards. In recent weeks, however, phones had appeared in many of the rooms, linked to a small exchange designed and installed by Eryl Wynn-Williams. So Walton was able to pick up a phone, dial the number for Cockcroft and tell him what was happening. In a few moments the older man was there and he quickly repeated the routine of tests – on, off, on, off. Cockcroft was no specialist in scintillation counting either but he had more experience than Walton and he believed that this was the real thing. They called Rutherford and once he had heard their story he naturally wanted to see the screen for himself. Always solidly built, the director had put on some weight in his later years and with his bad knee was neither fit nor supple; the little hut had not been built to accommodate such a person. Since they were able to turn the voltage off he was spared the business of crawling on the floor, but some sacrifice of dignity was required as he was squeezed into the hut. He made no complaint and once he was settled the two researchers retreated to the control table and turned on the apparatus. Soon the quarterdeck voice began to boom out, 'Switch off the proton current!', 'Increase the accelerating voltage!' and other such commands.[7] After some more of this he eventually called for the machine to be shut down again and once it was safe he emerged and made himself more comfortable on a lab stool. 'Those scintillations look mighty like alpha particle ones to me,' he declared, adding for emphasis: 'I should know an alpha particle scintillation when I see one, for I was in at the birth of the alpha particle and I have been observing them ever since.'[8] Chadwick was summoned next and he concurred.

What was going on? The stream of projectiles rushing down the acceleration tube at speeds of several thousand kilometres per second was, as we know, composed of protons, while the target into which they were crashing was made of lithium. What the four men had seen through the microscope was definitely not more protons, for ample precautions had been taken to filter them out. As Rutherford knew well, they were alpha particle flashes. And how alpha particles might emerge from a collision of protons with lithium was a piece of arithmetic they could all do in their heads without conscious thought. A proton, being the nucleus of a hydrogen atom, has a weight of 1, while a standard lithium nucleus has a weight of 7. Combine the two and you have an entity of weight 8, which is precisely equivalent to two alpha particles each of weight 4. Whether any of those present was yet prepared to speak the words out loud is not known but the implication was crystal clear: Cockcroft and Walton had not only penetrated the nucleus of lithium, they had split it in two. No comparable nuclear reaction – nothing remotely like it – had ever been observed before. All of the early Rutherford-Chadwick disintegrations had involved alpha particle projectiles causing the expulsion of single protons. The polonium-beryllium effect observed by Bothe, Becker, Joliot and Curie, and then explained by Chadwick, had also relied on alpha particle projectiles and involved the ejection of single neutrons. This new reaction was very different from either. They were putting in protons, the lithium was essentially vanishing and what emerged were two alpha particles – nuclei of helium. They were not merely chipping bits off a nucleus; they had gone right to its heart.

No less remarkable than the effect was the means by which they were achieving it, since this meant that the artificial acceleration of particles had actually worked. For the first time, a man-made apparatus had penetrated the nucleus, and better than that shattered it. The machine they were operating had actually achieved something that the natural radiations from radium and polonium had been unable to do. After three years and four months of work here finally was the laboratory tool that Rutherford had called for in 1927, and it did just what he had asked. Most remarkably of all, it had done so at voltage levels even lower than those predicted by Cockcroft on the basis of Gamow's work, back when they began. Although almost everybody had doubted those predictions at some time, here they were breaking up lithium with less than 300,000 volts, in fact by the look of it less than 200,000 volts. A million volts were not needed, still less the 2 or 3 million that most physicists had expected or the 10 million sometimes spoken of.

It was Rutherford who had the longest view on this, a view that stretched all the way back to 1909, to the day in Manchester when a student who was hardly more than a boy accosted him on the stairs and provided him with the evidence of something strange and solid inside the atom. Strange and solid the nucleus had remained ever since, but above all it had been elusive. The more he had learned in the intervening years about that tiny fly buzzing about in the huge, empty cathedral of the atom the more tantalizing it had become. He had weighed it, measured it, prodded it and chipped it but he had never had the means to get inside it to see how it worked. Now, on this spring morning in 1932, with that unexpected splash of sparks on Walton's screen, what had been his biggest, boldest and almost certainly, had it failed, his last attempt to penetrate the nucleus had borne fruit, and even the very first results took the breath away. This clumsy machine, he knew already, was not the end of it – others more powerful would soon take the work forward – but beyond all doubt, after twenty-three years, it was a new beginning.

There and then, in the midst of the excitement as he sat with Chadwick, Cockcroft and Walton, Rutherford marked an extraordinary occasion with some extraordinary decisions. Thorough checks were obviously required to confirm what they had observed and these should begin immediately, he said. They should be completed by the weekend so that a letter could be sent to *Nature* for inclusion in the following Saturday's edition, nine days off. As soon as that was done Cockcroft and Walton must throw themselves into an intensive round of experiments, applying their beam to as many elements as possible and recording the consequences using the most sophisticated means to be found in the Cavendish. Other laboratories would be quick to follow this work, and that meant not only Washington and Berkeley, Caltech and Berlin, but others too – so low were the voltages required that dozens of labs around the world were capable of matching what the Cavendish had done. It was vital, therefore, that they make the most of their head start, and to that end Rutherford ruled that the two men would be permitted to work day and night, with the laboratory opening specially for them after 6 p.m. In the meantime, until the letter appeared in *Nature*, he also decreed that none of the four should reveal what had occurred to anyone, not even their colleagues in the laboratory.

The rush to publish was to be expected; this was one of those moments when scientists set aside the rational and race for print. Lawrence, and possibly Tuve too, might be conducting similar experiments in that very week; any delay beyond the minimum required to ensure their observations were correct might mean that Cockcroft and Walton, their laboratory and its director would be denied the credit for being first. It was a matter of pride, and of history. That they should be permitted to work at night to make the most of their discovery was a rarity but not quite unheard of, as the Chadwick experiments had shown. But the secrecy was another matter and there appears to be no other instance of Rutherford behaving in quite this way. Chadwick had been discreet about the neutron it is true, but there had been no policy of silence and he had shared the news freely with some senior colleagues.

That afternoon Cockcroft and Walton got down to work, conducting two series of tests of which the first was designed to measure the effect at different voltages. The scintillation screen was a crude tool for this and above 300,000 volts the number of flashes was so great that the screen simply glittered continuously all over and optical counting was out of the question. But higher voltages could be left for another day, to be dealt with using Wynn-Williams's equipment. One of the chief surprises was that low voltages could still produce results and the old-fashioned screen was quite capable of confirming that. So the familiar ritual began. The curtain was drawn over the open side of the observation hut, the black blinds on the laboratory windows came into use, the counter sat in darkness for twenty minutes and he was allowed only short stints at the microscope. These procedures were passing out of use at the Cavendish by 1932 and it is an irony that this, the first successful use of a particle accelerator, should also have been among the last major experiments to make use of the zinc sulphide technique. Starting when the screen was a steady glow of flashes they edged the voltage downwards to the point where the first rough count could be made. This was 252,000 volts, which gave a figure of more than 160 scintillations per minute. Usually any count over 100 per minute would not be relied upon but this was not a case where a few flashes here or there might make a difference; an estimate would do. Next the voltage was dropped to 230,000 volts and here the count was twenty-one flashes per minute. It was a very sudden drop in effectiveness, but it was still remarkable that disintegrations were occurring at all. Creeping down further they found that at 160,000 volts they could still count nine scintillations per minute. And they got their lowest figure at 125,000 volts, with an average of 2.4 flashes per minute, after which the flashes became too spasmodic to be worth recording.

It was as if the nucleus, hitherto so secret and silent, had crept up behind them and shouted 'Boo!' No one, not even George Gamow, had imagined that nuclei might be disintegrated at 125,000 volts. Over the past decade every physicist working in the field had grown accustomed to the idea that the protective shield around the nucleus was extremely powerful. Even the alpha particles from radium could only rarely breach it and their energy was equivalent to acceleration by 7 million volts. In the case of polonium alpha particles the figure was 5 million volts. How could it possibly happen that particles imbued with just one-fortieth of that energy were passing through the shield? The first point, as Cockcroft had observed in 1928, was that protons were better projectiles than alpha particles because of their smaller electrical charge. Alpha particles carried a charge of +2, and since like charges repel they needed to travel at exceptionally high speeds to overcome the positive electrical barrier around the nucleus. Protons, with a charge of just +1, were bound to find the task easier. The second point, as Walton would observe later, was that this was a case where quantity made up for quality: the accelerated protons could not match radium alpha particles for speed, but there were far, far more of them in the beam – far more even than polonium could produce. Gamow had pointed out in the beginning, when he drew on quantum mechanical explanations for his theory, that the ability of particles to tunnel through the barrier was a matter of probability. Because the supply of particles from radium sources was so grudging the number of penetrations that could be achieved with them was always small. Cockcroft and Walton, however, had packed their beam with a veritable torrent of protons – something like 100 million million per second – and so the odds of success, even at low voltages, had tipped in their favour. It seems simple and logical now but the outcome was still a cause

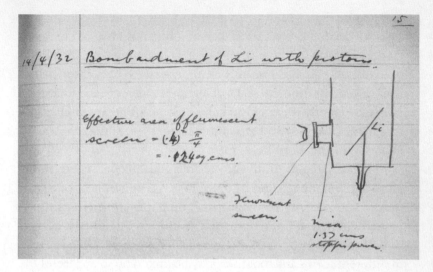

From Walton's notebook entry of 14 April: the arrangement
of the target, scintillation screen and microscope

of wonder to the four men who were aware of the secret.

A second series of tests that afternoon and evening established the energy of the alpha particles emerging from the
reaction. These tests were similar to the measurements carried
out by Chadwick in his neutron experiments and involved
placing metal foils of varying thicknesses in the path of the
particles to find the minimum needed to stop them. After
some calculations what emerged was another figure of great
importance: 8 million electron volts. On the face of it this too
was scarcely believable. A single proton accelerated by means
of 125,000 volts managed to tunnel through the protective
barrier of a lithium nucleus and lodged inside; immediately
the nucleus was violently destabilized and its components
resolved themselves into two alpha particles; having like
charges and being pressed close together, they repelled each
other with enormous force, each departing as if accelerated

using more than 8 million volts – enough to project it through a gas for several centimetres. The combined release of energy was more than 16 million electron volts, which meant that the relatively feeble proton had precipitated a reaction that was, given the minute scale of these events, extraordinarily violent. Happily there existed an equation of great simplicity to account for this and it was, even then, the most famous equation in the world: $E = mc^2$.

Einstein's description of the relationship between the energy of a body (E) and its mass (m) was the result of one of history's great imaginative leaps. (The letter c represents the velocity of light.) To the non-scientist it can be difficult even now to accept the idea that mass and energy are the same thing in different forms and that they may be linked by so apparently modest a mathematical device, but in 1932 it was already scientific orthodoxy, so when Cockcroft and Walton needed to account for the vast energy release they had witnessed they knew instantly that they were seeing the equation at work. All they had to do was add together the known weights of what they had started with, a proton and a lithium nucleus, and subtract the weight of what they finished with, two alpha particles. This gave a difference of just under 0.02 units of atomic weight, a small fraction of the whole but none the less a known quantity that had, as it were, gone missing. The lost mass did not manifest itself in any material particle or any ray but it could not have simply vanished because no such thing is possible. Instead it became energy. When the alpha particles flung themselves in opposite directions the energy they expended was that missing mass expressed in a different form, and Cockcroft and Walton were able to show that the quantities of mass and energy involved in the transaction accorded very neatly with $E = mc^2$. It was the first conclusive proof of the equation in a laboratory.

C. P. Snow, a frequent visitor to the Cavendish at this time, later wrote that when the two men finally turned off the machine and shut up shop at 10 p.m. that night, 'Cockcroft walked with sure-footed games-player's tread through the streets of Cambridge and announced to strangers, "We've split the atom. We've split the atom."' It was, said Snow, 'about the only magniloquent gesture of a singularly modest and self-effacing life'.[9] Sadly it is most unlikely that anything of the kind occurred – as Cockcroft himself said, Snow's idea was 'much more fancy than fact'.[10] It was too far out of character, as it would have been for Walton, and besides, Cambridge was a small place where gossip travelled fast, so it would have been a breach of Rutherford's requirement for silence. Perhaps their excitement cost them some sleep that night, but outwardly all remained as normal.

The next day, Friday, Cockcroft had an engagement in London and so did not appear in the laboratory. Instead Walton found a willing assistant in Rutherford, who wedged himself back into the hut and counted scintillations on and off for some hours while Walton continued to fiddle with the voltage. Some time that day, however, the moment came to get rid of the zinc sulphide screen and raise the blinds. They had done all they could with that technique; now they needed to confirm their findings by showing that the effect could be detected by the use of some other observational method. For this task they turned to the cloud chamber. Invented by C. T. R. Wilson twenty years earlier, this was the most celebrated of all scientific instruments to have emerged from the Cavendish. The idea was born when Wilson, a quiet Scot, started recreating clouds in a laboratory vessel, a sudden drop of pressure causing water vapour to condense into droplets. As in natural clouds the droplets formed around dust specks, but when Wilson removed the dust from his vessel he

discovered that they also formed around any available ions –
those broken atoms which carry an electrical charge. So he
made a shallow chamber in which he passed fast radioactive
particles through his man-made clouds and, lo and behold,
the particles created trails of ions and the droplets condensed
around them. Not only could you see the tracks but if you were
quick you could photograph them; they were like footprints in
snow. Even among seasoned experimenters these streaks of
tiny droplets never lost their power to amaze – Rutherford,
whose sense of the nucleus was strongly visual, was incapable
of using a cloud chamber without remarking on the beauty of
its images. It was not, however, a tool that could be used for
counting particles because the cloud effect was so shortlived.
Its value was mainly qualitative; the photographs provided
information about individual particle events.

In 1932 there were probably a dozen cloud chambers in
the Cavendish and Cockcroft and Walton found a basic,
mid-1920s model lying unused in some corner of the building.
This had to be connected to the experimental chamber of
their accelerator in such a way that the alpha particles produced
in the lithium reaction would enter the cloud chamber and
leave their tell-tale ionization trails. Another late night was
required for the fitting but on the Saturday the puffing and
wheezing of the cloud chamber and the clicking of its camera
added themselves to all the other drumming, clanking and
hissing sounds filling the big room. By 9 p.m. that evening
they had what they wanted: a set of photographs showing
clear tracks undoubtedly left by alpha particles, travelling a
distance which closely matched their earlier calculations. Two
independent methods had produced the same conclusion, and
no doubt whatever remained.

They rushed off to Rutherford's home, presented their
findings and plunged into a discussion of the implications.

Then, with the professor's help, they wrote a five-paragraph letter to *Nature* entitled 'Disintegration of Lithium by Swift Protons'. It began by mentioning their previous account of the apparatus and explaining that since that had been written they had inserted a target of lithium alongside a scintillation screen. When the voltage was applied, they wrote, bright scintillations were observed on the screen and these increased in number as the voltage was raised. The particles causing this effect also appeared when a cloud chamber was used, and their ranges were measured by two different means. All the findings pointed towards the same conclusion:

The brightness of the scintillations and the density of the tracks observed in the expansion chamber suggest that the particles are normal alpha particles. If this point of view turns out to be correct, it seems not unlikely that the lithium isotope of mass 7 occasionally captures a proton and the resulting nucleus of mass 8 breaks into two alpha particles, each of mass four and each with an energy of eight million electron volts. The evolution of energy on this view is about sixteen million electron volts per disintegration, agreeing approximately with that to be expected from the decrease of atomic mass involved in such a disintegration.[11]

Then a final sentence: 'Experiments are in progress to determine the effect on other elements when bombarded by a stream of swift protons and other particles.'[12]

This labour complete, they retired to their homes to pass a quiet Sunday and in the evening Walton wrote another letter, this time his customary one to Freda. It was a long one by his standards, running to four pages, and while it began with some uneasy small talk – he still had not really acquired the knack – it soon got down to business. 'Last Thursday was a red letter day for me,' he declared. 'Not only did I get a letter from you but Cockcroft and I made what is in all probability a very important discovery in the lab.'[13] He recounted how they had been able to 'smash up' nuclei and produce some interesting rays, and when they demonstrated this to Rutherford he became very excited, suggesting it be kept a 'dead secret'. ('I don't expect you will pass it any further,' Walton observed delicately to Freda.) He then gave an explanation for the secrecy:

He [Rutherford] suggested this course because he was afraid that the news would spread like wild fire through the physics labs of the world and it was important that no lurid accounts should appear in the daily papers etc before we had published our own account of it. It makes matters very awkward for me in the Cavendish as people are continually asking how our experiments are getting on and it is difficult to satisfy some of them with evasive replies.

Later in the letter, after he had described his three days of hard work and his many consultations with Rutherford, Walton mentioned another of the professor's concerns:

We will work as hard as we can next week as Rutherford is very anxious that we should get the credit for any other interesting discoveries in this field before other people start to get a move on when they hear of our experiments. We know that people in the States are working along similar lines and Rutherford would like to see any credit going to the Cavendish. He is not fond of American physicists in general on account of their tendency to do a great deal of boasting about very little.

You can almost hear the voice of Rutherford in the back-ground, blustering on about Americans and what they get up to. In fact he counted several Americans among his friends and had never been slow to praise American science, but his competitive edge was showing through. Lawrence, Tuve and Lauritsen all had the capability of repeating these experiments, perhaps within weeks; he did not want to have to share any more glory with them than he had to.

Freda's reply brimmed with warmth: 'My very heartiest congratulations on your great success in the lab. You must be feeling very excited, hugging your secret close until Saturday, when I'm sure you will be making a name for yourself and the Cavendish in very wide circles.'[14] She was impressed that her Ernest should have been working shoulder to shoulder with the great Rutherford and she confidently predicted that the newspapers would be very interested in it all. 'I don't think I can explain to you,' she wrote, 'how glad I am that it is you who has done this.'[15]

For some reason − perhaps Rutherford misunderstood *Nature*'s deadlines − the letter announcing the discovery arrived too late to appear in the following weekend's edition of the magazine and so there was an extra week's delay. This was put to good use, as Cockcroft and Walton had time to test their beam on other elements, starting with boron and fluorine. These were thought to be the most likely to disintegrate under proton bombardment, and sure enough when everything was set up and the beam turned on they shattered in much the same way that lithium had. A proton of weight 1 entered a boron nucleus of weight 11 and the result was three fast alpha particles, each of weight 4. A proton entered a fluorine nucleus of weight 19 and out came one alpha particle of weight 4 and an oxygen nucleus of weight 16. It was thrilling. By now the two scientists had installed a Wynn-Williams counter so that they could record their results on photographic strips, and after a long day in the lab Walton would rush home with the strips, often when they were still wet from the developing fluid, and work his way through them into the small hours, adding up and tabulating the 'kicks'. These he would type up in duplicate and bring to the lab in the morning to be discussed with Cockcroft and Rutherford. The director, for his part, could hardly keep out of the laboratory, so great was his excitement, and if he could not be present he insisted on frequent phone calls to keep him up to date.

It is no surprise in the circumstances that the secret could not long be contained inside the walls of the former Lecture Room D. Harrie Massey, a young theorist, picked up the story in cloak-and-dagger style one afternoon after tea when his friend Nevill Mott came over and asked: 'Have you heard that Cockcroft and Walton have done this?'[16] Rather than explain out loud, the lanky Mott leant over and wrote simply: $Li + H \rightarrow 2He$. Lithium plus hydrogen yields two helium; no further words were required. A spreading euphoria in the laboratory soon made a mockery of the secret, and word was also travelling further afield. Walton had leaked the news by telling Freda, although he knew that she was both discreet and well removed from the scientific mainstream. Cockcroft had naturally told Elizabeth too, and he took a greater risk in passing the word to Metropolitan-Vickers, although given the company's contribution it was only fair to do so. He wrote separately, it seems, to Allibone and McKerrow, swearing each to secrecy, and the result was a comical encounter between the two at the Manchester lab in which they dodged around each other for some time before realizing they were both in the know. They were delighted. Rutherford, meanwhile, wrote to Bohr, to whom he had only recently passed on the news of Chadwick's great discovery.

It never rains but it pours, and I have another interesting development to tell you about of which a short account should appear in *Nature* next week. You know that we have a High Tension Laboratory where steady d.c. voltages can be readily obtained up to 600,000 volts or more. They have recently been examining the effects of a bombardment of light elements by protons . . .[17]

He described with evident delight the experiments and findings, declared himself 'pleased that the energy and expense in getting high potentials has been rewarded' and concluded: 'You can easily appreciate that these results may open up a wide line of research in transmutations generally.'

Bohr replied:

By your kind letter with the information about the wonderful new results arrived at in your laboratory you made me a very great pleasure indeed. Progress in the field of the nuclear constitution is at the moment really so rapid, that one wonders what the next post will bring, and the enthusiasm of which every line in your letter tells will surely be common to all physicists. One sees a broad new avenue opened, and it will soon be possible to predict the behaviour of any nucleus under given circumstances.[18]

Before Bohr's letter reached him Rutherford had decided not to wait for Cockcroft and Walton's communication to appear in *Nature* because a perfect opportunity to make the announcement presented itself two days earlier, on Thursday 28 April. That evening he was due to chair a meeting at the Royal Society in London on 'The Structure of Atomic Nuclei'. This had been organized so that physicists would have an opportunity to discuss Chadwick's neutron but Rutherford could see that it would be invidious to sit through such a gathering without mentioning the latest discovery too.

Furthermore, with all the leading British atomic physicists due to attend, it would have been hard to imagine a better forum for the announcement. He therefore arranged for Cockcroft and Walton to be present.

It was a fine piece of theatre. Rutherford opened the proceedings and for perhaps fifteen minutes followed his prepared text, supplying, as was his custom, an authoritative *tour d'horizon*. When he came to the discovery of the neutron, of which everyone present was aware, he limited himself to setting out the background up to the point where a 'new type of radiation' was found that demonstrated 'surprising properties', and then said gracefully that he would leave it to Chadwick, the next speaker, to explain his work on this phenomenon.[19] Next, still sticking to his text, he pressed on to a conclusion that noted with satisfaction the 'comparatively rapid progress' achieved in recent years in the acceleration of particles. Naming Tuve and his colleagues, Lawrence and Livingston and Cockcroft and Walton, he spoke of a 'hopeful prospect' that high-speed particles might soon be put to use in nuclear experiments.

This was the end of the script circulated in advance and no doubt the assembled Fellows now looked up in the expectation of seeing Chadwick step to the rostrum. But Rutherford stayed where he was, announcing that he had something to add. In recent weeks, he revealed, Cockcroft and Walton had conducted some 'interesting new experiments' in Cambridge. He explained first, in terms very close to those of the *Nature* letter (which no one present had yet seen), that they had succeeded in disintegrating lithium at 125,000 volts and had observed the emergence of alpha particles with energies of 8 million electron volts. And then he jumped beyond the *Nature* announcement, stating that by that date they had also disintegrated boron, fluorine and aluminium and all had thrown out alpha particles. Some scintillations had also been observed from beryllium, nitrogen and carbon, although their nature was not yet clear. Many questions were raised, said Rutherford, and there was much work to be done but it was evident that a 'new and wide field of research' had been opened up. With that he declared: 'Dr Cockcroft and Dr Walton are to be congratulated on their success in these new experiments which have taken several years of hard work in preparation.'[20] A gesture of the hand brought the two men to their feet to acknowledge 'a good clapping', as an embarrassed Walton called it, from an audience equipped as few others would be to recognize the scale of their achievement. Afterwards toasts took the place of applause when they were the guests of Rutherford and Fowler at a celebratory nine-course session of the Royal Society Dining Club. It was, as a colleague would later write, 'a very great day for them all'.[21]

14. Still Safe

The senior staff of the Cavendish Laboratory had some experience of news management. As Walton mentioned, Rutherford was concerned from the start to prevent 'lurid accounts' appearing in the newspapers, and that was one reason for his temporary rule of secrecy. Though the director believed passionately in the need for scientists to communicate with the public he had been at the centre of events long enough to learn the distorting effects that journalistic competition and ignorance could have. This was one reason why he had been happy to encourage J. G. Crowther, who arrived on the scene at the end of the 1920s as the first genuine science correspondent in the British press. Crowther had attended the Cavendish as a research student briefly in 1919, only to drop out when he suffered a nervous breakdown. After a spell as a school science teacher he took a part-time job with the Oxford University Press and then, showing great determination and enterprise, forced his way into journalism.

A few newspapers at that time retained working scientists as consultants, ready to write about new developments, but Crowther persuaded the *Manchester Guardian* that the way of the future was dedicated specialist journalism. First as an occasional contributor and soon as a regular, he showed that he could find stories as well as react to them and that he could write about science in a way that readers understood. And just as C. P. Scott, the paper's editor, became convinced of Crowther's value so did Rutherford, who allowed the journalist to become a frequent visitor to the Cavendish, on friendly terms with several of the scientists. When Chadwick first addressed his colleagues on the discovery of the neutron Crowther was one of the few outsiders in the audience and was consequently the first journalist to report the news. His article was enthusiastic but also informed and intelligent, and as Chadwick and Rutherford had hoped it set the tone for the coverage elsewhere.

Although only a few weeks had passed since then it proved impossible to repeat this feat for Cockcroft and Walton because Crowther was abroad, pursuing the atomic story in Copenhagen. He had been lured away to attend the spring gathering at the Niels Bohr Institute, where the neutron was top of the agenda. Arriving there with the Cavendish seal of approval, he was allowed to attend some of the discussions. Heisenberg was present – 'socially charming: what a man he is!'[1] – as was Dirac, but it was Bohr himself who most impressed the journalist. 'No one could have fewer obvious qualities of leadership,' he wrote. 'Extremely diffident and verbally diffuse, he jumps up and says what is in his mind, and so do others, yet such is the respect for him that ultimately order prevails. Young men had played ping-pong when he wanted to start, but he did not say anything and in a minute or two they stopped.' After the symposium Bohr invited Crowther to tea with the institute staff, an occasion largely given over to the latest Dirac stories. (An example: when Bohr took Dirac to an Impressionist exhibition and asked him which of two paintings he preferred, the Englishman pointed at one and declared: 'I like that, because the degree of inaccuracy is the same all over.') And once he had finished with high physics, Crowther took himself off around Denmark to investigate the state of science there and study agricultural developments.

All of this meant that he could not be present at the Royal Society to hear Rutherford announce the success of Cockcroft and Walton and as a result the news broke in a most disorderly fashion, emerging first in a popular Sunday paper called *Reynolds's Illustrated News*. Although it was on that same weekend that the letter to *Nature* finally made its appearance, the periodical was not the paper's source. The tip-off appears to have come a day earlier from someone who attended the Royal Society discussion, and this gave the editor time to make the most of his scoop. On the morning of Sunday 1 May, therefore, overshadowing items such as 'Peeress Robbed On Road' and 'Canal Bursts Its Banks', a single headline straddled the whole front page blaring the words: 'SCIENCE'S GREATEST DISCOVERY'.

The first paragraph gives a flavour of the reporting:

A dream of scientists has been realised. The atom has been split, and the limitless energy thus released may transform civilisation. On the authority of Lord Rutherford, the world-famous scientist, *Reynolds's* is able to announce exclusively that years of patient experiment at the Cavendish Laboratory at Cambridge have at last been successful. The effect of splitting the atom is that the electrical power now available to mankind may be multiplied 160 times. This is the greatest scientific discovery of the age.

There was no reference to the *Nature* report, which was so different in tone, and no mention, for that matter, of the words 'nucleus', 'lithium' or 'accelerator'. The only source acknowledged was Rutherford himself: cornered by a *Reynolds's* reporter at a public engagement in London the previous day, he had scribbled on a draft of the story a curt confirmation that the experiment had taken place and the comment: 'Your conclusion is fairly correct.'[2] It is difficult to believe that even this guarded endorsement applied to the report as printed.

★

How the story broke

Walton's landlady at Albion Row in Cambridge was not a reader of *Reynolds's News* but her own paper, the *Sunday Express*, had managed to scramble together a short response to its rival's scoop. When she reached page 15, therefore, she was able to read the cryptic announcement: 'The Atom Split, But World Still Safe'. And there, in the middle of the text, was the name Dr E. T. S. Walton. Quickly she alerted her lodger to his first taste of fame. At the Cockcroft house in Sedley Taylor Road, meanwhile, the telephone was soon ringing as reporters for the Monday editions began following up the story, and by the evening this British scientific triumph was featuring on the BBC radio news reports.

It was not until they turned up at the laboratory on Monday morning, however, that Cockcroft and Walton felt the full force of what had been unleashed, for not only was a small crowd of journalists waiting to ask them questions but there were also photographers demanding pictures of the new heroes at their apparatus. Walton was driven 'nearly demented'[3] by the interrogations until a horrified Rutherford, who had already banned the photographers from the lab, announced there was nothing more to be said. Eventually, when calm returned, he was persuaded to pose briefly with his two protégés. The result is a priceless image: Rutherford almost dapper in grey Homburg hat and dark suit with wing collar and watch-chain, Cockcroft with heavy bags under his eyes but sporting a racy striped necktie and respectable three-piece, and Walton with the pockets of his shapeless tweed jacket bulging to the point where the single button is aching at the strain. All three wear expressions that combine happiness with embarrassment in roughly equal measure.

It was not just the attention that made them uncomfortable but the content of the reports, which often bore little relation to their own ideas about their work. For a start, those words 'splitting the atom' were at best an approximation of what they had done, and certainly in need of elaboration. Breaking up atoms was old hat: scientists had been knocking electrons out of them for many years. The novelty, of course, was to split *nuclei* effectively, but unfortunately 'nucleus' was not a familiar enough word to be used in headlines. Much more unsettling than the imprecise language, however, were the suggestions being made about the implications of what Cockcroft and Walton had done. The *Daily Mail* was one of those to treat it as a piece of alchemy, declaring: 'It opens up new horizons and perhaps in time will render practicable the transmutation of lead into gold.'[4] Again there was a germ of truth in this – they had, after all, changed lithium into helium – but the idea that such machinery might ever be employed to manufacture gold in a quantity that was visible, let alone useful or valuable, was preposterous. More common still in the newspaper reports was the line taken by *Reynolds's*, that the Cavendish had tapped into a source of energy powerful beyond the imaginings of the people of 1932. This single scientific event, it was suggested, might herald an age of universal comfort and prosperity – or it might mean apocalyptic destruction. As the *Daily Mirror* warned: 'Let it be split, so long as it does not explode.'[5]

A coincidence encouraged such notions: at precisely the time that the story broke the British newspapers were also carrying reviews of a play called *Wings Over Europe*, just opening in the West End. Written by Robert Nichols and Maurice Browne, this was the tale of a young scientist who finds the formula for controlling the energy in the atom, demonstrates his power by detonating a lump of sugar and leaving a crater as big as Vesuvius, and then tells the British government to make plans for the international control and exploitation of this priceless resource. When, for want of vision, they refuse, he resorts to blackmail 'At one o'clock tomorrow, England ends,' he declares. 'Where this island was, will be a whirlpool of disintegrating atoms.'[6] At the climax he is shot and catastrophe averted, but the story does not end there. As the curtain falls a message arrives from the 'United Scientists of the World' announcing that they too have the formula and that atomic bombs will rain on the capitals of the world unless a congress is convened to ensure the wise use of the new energy source.

Although it had enjoyed success on Broadway the production did not find much favour with London critics, who disliked its melodramatic tone and stereotypical British characters. None the less it was one of those plays whose subject matter ensures it attention, as its authors intended. Nichols and Browne called it 'a dramatic extravaganza on a pressing theme' and, topping up the verisimilitude, insisted that their hero was 'only a concentration into dramatic form of the general powers and claims of science today'.[7] Since the play opened just days before the news from Cambridge reached the headlines, the two events inevitably became intertwined. Newspapers were unable to resist discussing the scientific breakthrough in terms of Nichols and Browne and the play in terms of Cockcroft and Walton, and though everyone at the Cavendish found this emphasis distasteful they were powerless to prevent it. In vain did they point out, again and again, that a fundamental misconception underlay the reporting. While it was true that in each reaction caused by their machine a proton accelerated by 125,000 volts unleashed other particles with a combined energy equivalent to acceleration by 16 million volts, that was not the whole story. Only about one proton in 10 million actually penetrated a target nucleus and since all the rest had equal energy, a vast amount more was wasted in the disintegration process than was created. As Rutherford put it, in energy terms the atom was still a sink rather than a fount.

Walton soon came to a conclusion reached by many people who suddenly find themselves in the news, reporting to Freda: 'I have learnt one thing from reading the papers this week, and that is not to believe all you see in them.'[8] She had no such concerns. 'I am having a lovely time this week collecting a picture gallery of you,' she wrote in delight,[9] noting that the news had been reported in the *Irish Times* and the *Irish Independent*, and that County Waterford, where Walton had been born and where she now worked, was claiming him as its own. For all her pride and pleasure, however, she appears to have hinted to no one that this Ernest Walton whose name was on everyone's lips was anything more to her than a vague acquaintance from schooldays. 'It was very funny,' she wrote, 'when coming out of church on Sunday night, talk turned to you and who you were among people in the porch, somebody asked me if I had by any chance been to Methody when you were there and I said I had (and remembered you!).'[10] She had written of him 'hugging his secret'; now it was her turn. Walton had heard from Methodist College, their old school in Belfast, that the flag had been flown and pupils granted a half-day holiday in his honour. This was one of many messages of congratulation, along with quite a few sent by 'lunatics giving me solemn warnings',[11] but he joked that it was only when he received a letter from his younger brother Jim that he knew something truly extraordinary must have happened. A glow of happiness rises from his letters in these weeks and in one of them, to Freda, he made a gesture as eloquent as it was overdue. 'I hope you will know who this letter is from,' he wrote at the foot of the page, 'if I omit the "W" after my Christian name. Yours, Ernest.'[12]

At the Cockcroft home letters also arrived by the fistful, many reporting high excitement in Todmorden. Aunts and uncles had leapt from their armchairs in delight when they heard the Sunday evening radio bulletin and scurried out into the streets to spread the news. Elizabeth's father declared that never before had a son of Todmorden caused such a splash and that John should have the Freedom of the town, while another relative predicted a knighthood and said they would have to get 't' band out'.[13] A third writer sent John warm congratulations and added: 'Your aunt would be less sceptical about the performance if you would send her half a proton so that she might put it in her reticule and at the right moments bring it out . . .'[14] John's mother, as mothers will, informed him that he should immediately take a rest or visit a doctor. Having seen the photograph of the three scientists in the paper she was most concerned at her boy's appearance – 'so much like your father at the last that it gave me quite a shock'.[15] And like many another in Todmorden she allowed herself the afterthought: 'I wonder if your income will increase now.'

On 4 May there was another occasion for Cockcroft and Walton to savour. Albert Einstein, then visiting Britain, came to Cambridge to deliver a lecture and paid a call at the Cavendish, where he was naturally brought to view the now-famous accelerator and to meet its architects. Walton explained the operation of the machine to this 'very interested' visitor, put it through its paces and answered a few questions. There followed in the evening a grand dinner in Einstein's honour at St John's College (Walton discovered that he did not like caviar) and later a private chat with the great man involving Eddington, Cockcroft, a handful of others and Walton. 'He seems a very nice sort of man. I felt as if I had been highly honoured,' he informed Freda.[16] Einstein too was impressed, writing a few days later of his 'astonishment and admiration'[17] at what he had seen at Rutherford's laboratory.

The names of the two men were travelling around the world (although in Cockcroft's case it was usually spelt 'Cockroft'). From Sweden to Italy and from Argentina to India newspaper readers were learning of the Cambridge scientists who had 'split the atom'. Movietone News even turned up at the Cavendish hoping to film the heroes with their apparatus but Rutherford would have none of it. The sensationalism and muddle about the implications of the breakthrough, more-over, refused to die down and Cockcroft for one became fed up. He vented his frustration on Crowther as soon as the hapless journalist returned from his Danish trip:

Dear Crowther,
You have let us down badly – we really relied on you to get all this business straight in the Press first so that we could

simply refer all other people to you. Instead of that, owing to your most unfortunate absence, *Reynolds's* newspaper got a very garbled account from someone in Bristol, and the rest of the Press followed suit. However one tries to put things right, short of writing an article oneself, one only seems to make it worse![18]

Not least of the confusions to have arisen was one in a report carried by the Associated Press, the leading American news agency, which suggested that British scientists had converted helium into hydrogen rather than lithium into helium. When this appeared in the US papers the claim was quickly queried by readers and the agency's embarrassed London bureau chief was obliged to cable Rutherford, Cockcroft and Walton begging for an interview to clarify matters. Pointing out that the AP service went out to 1,400 newspapers across North America, he promised that 'we have in mind nothing sensational'[19] and that he would be more than happy to submit the text to them for approval. Cockcroft, seeing a chance for Crowther to make good his failure, arranged for him to write the piece, which was subsequently wired to New York. There, the story was as much a sensation as it had been in Britain. The *New York Times* carried three lengthy items in a week, including a front-page report and a leading article that praised the work as a major scientific advance and dismissed talk of nuclear energy as premature. In the paper's Sunday edition came a full-page feature: 'The Atom Is Giving Up Its Mighty Secrets'. Describing in some detail the experiment and its outcome, it stated: 'Never was a result more unexpected obtained.'[20]

It has been said that when Ernest Lawrence heard the news from England he broke off his honeymoon to send an urgent cable to his lab in Berkeley, but the facts do not quite bear this out. It is not that he was above mixing physics with romance (even after his marriage he liked to visit interesting laboratories while on holiday, leaving his wife to amuse herself) but that the chronology makes it almost impossible. Lawrence married Mary Blumer in New Haven, Connecticut, on Saturday 14 May 1932 and it seems they departed promptly for a honeymoon on Lake Champlain in Vermont. But the news of the breakthrough in England had reached the United States almost two weeks before the wedding and the confusion caused by the first Associated Press report had been cleared up by 8 May at the latest – the day of the big *New York Times* feature. It is difficult to imagine a scientist as dedicated as Lawrence waiting around before his marriage in a well-informed university town such as New Haven and failing to notice such a development. And it is all the more difficult to imagine him departing on his honeymoon in a state of ignorance when we know that the guests at the wedding included several of America's leading physicists, including at least two – Merle Tuve and Cockcroft's friend Joseph Boyce – who had a direct interest in the story. The groom must have known and he must have discussed it with his friends.

Whatever the circumstances in which he heard, he is said to have been 'momentarily disturbed'[21] at the news, a natural response. It was not so much that his cyclotron had been beaten to this milestone, although he probably did not savour the feeling, but that 125,000 volts was such an extraordinarily low figure. Very soon, however, he felt the surge of exhilaration that befitted a man who liked to tell students: 'There are discoveries enough for all.'[22] It was then that he cabled Berkeley with the message remembered by a colleague: 'Get lithium from chemistry department and start preparations to repeat with cyclotron.'[23] In fact it would not be quite so easy as that, as Stanley Livingston noted: 'Well, we weren't ready for experiments yet. We didn't have the instruments for detection. I had built the machine but had not included any devices for studying disintegrations. So we had to rebuild it.'[24]

As for Merle Tuve, he would look back half a century later and admit: 'They caught us with our pants down.'[25] Of all those working on particle acceleration he had been at it longest and yet when the Cambridge pair announced their success he still did not have an apparatus capable of matching the results. At the time the news broke he and his team were preparing to erect a big Van de Graaff machine with two-metre spheres outside on the lawn, for want of proper accommodation. Soon complete, it proved alarming to passers-by, as dust and insects in the vicinity of the spheres tended to cause impressive sparking, but it ran successfully up to 2 million volts and the team found that their tube could withstand 1 million. There was no question, however, of conducting disintegration experiments in the open air so the generator was taken down until a home could be found for it. While Tuve fought that battle the team began building a more modest 600,000-volt machine indoors – a machine of that size, they now knew, would be useful after all.

In all the leading labs people were pinching themselves, for the astonishing truth was that, with disintegration occurring at 125,000 volts, any one of them might have achieved it long before. Lawrence and Livingston had a cyclotron capable of that voltage in mid-1931; Lauritsen might have done it with the apparatus he built in 1928 and Tuve, had he known the trifling power supply needed, could surely have begged or borrowed the necessary equipment even before that. But the people pinching themselves hardest were almost certainly Cockcroft and Walton, who had actually attempted disintegration experiments on lithium both the previous year, in

April 1931, and a year earlier, in June 1930. In both cases they had applied a more than sufficient voltage and the reaction undoubtedly took place, but they had missed it because they were looking for the wrong thing. Cockcroft would say a few years later: 'The facts are that we looked first for gamma rays and not alpha particles, since at that time we had a fixed idea that gamma rays would be the most likely disintegration products.'[26] If they had put a scintillation screen beside their lithium target in the summer of 1930, therefore, they would have made the same discovery with their first apparatus down in the basement room. The move to Lecture Room D, the construction of the voltage multiplier and the erection of the rectifier tower and cylindrical acceleration tube had all been, strictly speaking, unnecessary.

At root, however, this was more than a matter of over-looking a scientific possibility. At Washington and Berkeley, just as much as at the Cavendish, physicists had been well aware by 1930–31 of the potential implications of Gamow's ideas. Tuve's colleague and adviser on theoretical matters, Gregory Breit, knew that disintegration might be possible at levels far below those of radium alpha particles and Stanley Livingston noted the same thing in his Ph.D. thesis in the summer of 1931. But a theoretical awareness was not the same as a belief, and nothing could quite displace in their minds the conviction that anything less than 1 or 2 million volts was simply too puny. In their conscious and even their subconscious minds the words were etched as if in stone: the atomic nucleus is not only tiny and elusive but it is also a fortress of unimaginable strength. Cockcroft and Walton were no different. In both 1930 and 1931 they did not bother to linger on experiments at 250,000 volts because in their hearts they did not believe it was nearly enough. Instead they rushed onwards to 700,000 volts, and even in the early months of

1932 they were already talking of raising the voltage multiplier capacity to a million volts. Older and wiser scientists shared this conviction, and not just Rutherford and the senior Cavendish staff but also all those visitors to Cambridge who saw the machine and discussed its workings with Cockcroft and Walton, among them Bohr and Millikan. That not one of these foresaw the lithium reaction at 125,000 volts is the very best proof of the enigmatic status of the atomic nucleus up to early 1932.

Why did the Cavendish get there first? In one sense the explanation is the story of this book, for it was the momentum that the whole laboratory had accumulated in the preceding five or six years that carried it across the line. Chadwick's discovery of the neutron was a similar case: the polonium-beryllium reaction had been observed by Bothe and Becker, by Joliot and Curie and by others too, but it was only at the Cavendish that it was understood. Frédéric Joliot was amazed and dismayed when the truth was revealed, but he remarked later that 'old laboratories with long traditions always have hidden riches'.[27] His own laboratory in Paris had plenty of heritage but that was not quite the same, for the Curies had never shared Rutherford's obsession with the nucleus. So when Chadwick smelt the neutron in that Joliot–Curie paper in *Comptes Rendus*, he was tapping into those hidden riches of the Cavendish: the first intuitions of Rutherford in 1920; the years of sometimes silly experiments they had carried out together; the many, many nights spent thinking about the neutron after the infamous 6 p.m. closure. Joliot confessed that, though the concept of the neutron had lain in the literature for more than a decade, he had never noticed it. How different from Rutherford's laboratory.

In similar fashion the accelerator project at Cambridge, although no grander in scale than its rivals, simply had more weight behind it. And above all it had Rutherford behind it. Even in his sixties, the battleship of physics was capable of a determination that few others could match. One who could was Lawrence, but while Lawrence's dynamism enabled him to push the cyclotron project forward at astonishing speed, he did not quite have the vision and impatience needed to make that first breakthrough. Rutherford was not burdened by any interest in machinery and never had been in his long career; all he wanted – as he never ceased to tell his students – was results. And even though, as his remarks at the Metro-Vick opening show, he had never really accepted the idea of disintegration at less than a million volts, he supported the project with his customary enthusiasm from start to finish. He told Cockcroft to go ahead and gave him resources; he gave him a partner in Walton and made sure that Walton was able to see the work through; he gave them space when they needed it and resources when they asked for them and, when the moment came, he pushed them into making the crucial experiment. Did he know what they would find? Almost certainly not, but he knew that the time had come to start looking.

And besides Rutherford himself there were many other riches at the Cavendish for Cockcroft and Walton to draw on, that were not available elsewhere. It was a laboratory devoted to the atom and to the nucleus within it. Whatever was known in this field – techniques, equipment, mathematical tools, even theory – it was known by someone there, and more than that it was discussed, challenged and tested at colloquia and other gatherings. To any problem or difficulty in atomic physics there would surely be an answer somewhere in the laboratory in Free School Lane. And for problems in engineering they were fortunate to have the support of Metropolitan-Vickers: without the transformer, the Burch pumps and the Compound

Q that came from the Manchester company, not to mention other support and advice, the whole enterprise would have taken far longer, if it had been possible at all. The big machine in the former Lecture Room D was not the most elegant or the most ingenious in its class, but because of the rich culture from which it sprang it did its job, and it did it first.

By the end of the year not only were Lawrence and Tuve getting results – some of them no less surprising than the proton-lithium reaction – but so were Lauritsen at Caltech and Brasch and Lange in Berlin, who had immediately seen the potential of this field and dropped their X-ray work. Indeed a dozen other laboratories were able to enter the field and dozens more were rushing to follow. Rutherford himself could not resist joining the fray and in great haste had Oliphant build a second accelerator in the Cavendish, employing more modest voltages but a more intense beam. Then, displaying much of the vigour of his youth, he plunged into another series of *bahnbrechend* experiments. And accelerators were not the only way forward. As the theorists and the experimentalists digested the implications of Chadwick's discovery, the neutron, they soon saw that it too would be a powerful tool for probing the nucleus. Having no electrical charge of its own but considerable mass, it could penetrate nuclei of heavier elements than those vulnerable to the Cockcroft-Walton machine or even the more powerful Berkeley cyclotron.

A further discovery came that summer from a different quarter. Carl Anderson at Caltech was among a small but growing band of physicists studying particles that came, not from atoms, but from outer space. These cosmic rays, as they were called, had great energies and Anderson was photographing them in a cloud chamber when he came upon an unexpected track. From this single track he was able to deduce the existence of another new particle, the positron. This was the positively charged counterpart of the electron. Having been first with the neutron and first with disintegration by accelerator, the Cavendish had to admit that it had been pipped at the post with the positron, for it was clear that this particle had also been cropping up in disintegration experiments in the Cambridge laboratory. Blackett had been working on the problem for a year with an Italian researcher, Giuseppe Occhialini, and they had accumulated hundreds of cloud-chamber photographs which suggested the presence of positrons, but Blackett had never been sufficiently confident to publish. Anderson jumped in first.

What was in effect a new field of science, nuclear physics, was being born. The phrase had existed for some time and a few scientists had already devoted years of study to the nucleus but what had been an obscure and often barren region now suddenly attracted a great throng of the finest people, both experimenters and theorists. The neutron and the positron, the new counters, the cyclotron, the Van de Graaff generator and the Cockcroft–Walton accelerator all offered such a range of possibilities that, as Lawrence had said, there were discoveries enough for everyone. One measure of the transformation can be found in the pages of *Physical Review*: in that first year of advances, 1932, 8 per cent of contributions related to nuclear physics; a year later it was 18 per cent and in 1937 32 per cent. Almost a third of all papers in this leading American journal, in other words, was devoted to this single aspect of physics. Another indicator lies in the editions of Gamow's book on nuclear theory: when he published it first in 1931 only a handful of nuclear reactions was known but by the time of the second edition in 1937 he could include a list of well over a hundred, each of which represented a new means of understanding the nucleus. Elusive for so long, the fly in the cathedral was at last becoming known.

15. Nobel

Almost twenty years after the Cavendish breakthrough, on the evening of 15 November 1951, Ernest Walton was at his home in Darty, Dublin, when a ring of the doorbell announced the arrival of a telegram. It came from Stockholm and it read simply: 'Swedish Royal Academy of Science has awarded you one half of the Nobel prize for physics 1951. Letter follows.'[1] After a long period of quiet this was the beginning of a frenzy of the kind he had experienced once before, with cheerful messages pouring in and newspapers begging him for interviews. The other half of the prize had naturally gone to Cockcroft, who wrote promptly: 'My dear Walton, I expect that you are feeling extremely thrilled as we are . . .'[2] Three weeks later they were both in Sweden to hear the grounds for the award set forth at the presentation ceremony:

By its stimulation of new theoretical and experimental advances, the work of Cockcroft and Walton displayed its fundamental importance. Indeed, this work may be said to have introduced a totally new epoch in nuclear research . . . The great nuclear scientist Rutherford, with whose work your discovery is closely connected, sometimes used to say: 'It is the first step that counts.' This saying may be applied in the truest sense to your discovery of the transmutations of atomic nuclei by artificially accelerated particles. Indeed, this work of yours opened up a new and fruitful field of research which was eagerly seized upon by scientific workers the world over. It has profoundly influenced the whole subsequent course of nuclear physics. It has been of decisive importance for the achievement of new insight into the properties of atomic nuclei, which could not even have been dreamt of before. Your work thus stands out as a landmark in the history of science.[3]

Cockcroft, responding at the banquet that followed the ceremony, explained modestly that they had been fortunate to work at the Cavendish at a time when new theoretical ideas and technological advances opened the way for discoveries. Walton, evidently emotional, said of the prize: 'It is an honour so great that, even yet, it is difficult for me to believe that it is true.'[4]

By 1951 neither worked at the Cavendish. After their first great success they stayed at the cutting edge of research for just two years, making further discoveries, publishing papers, engaging in scientific controversy and touching a high point in October 1933 when they were invited to the Solvay conference, a physics summit meeting in Brussels. There can have been few more distinguished gatherings in all the history of science: Rutherford, Marie Curie, Langevin, de Broglie, Joffe and Bohr were among the old guard; Heisenberg, Schrödinger, Dirac, Gamow, Mott and Pauli spoke for the new generation of theorists; Chadwick, Bothe, Meitner, Enrico Fermi, Blackett, Lawrence, Joliot and Irène Curie represented the experimenters. Half of the forty men and women present had won or would win the Nobel prize. Yet within a year of that occasion both Cockcroft and Walton were slipping out of the mainstream.

Cockcroft remained at the Cavendish until the war but concerned himself increasingly with administration, in particular the running of the magnetic laboratory, which he took over altogether in 1934. With the approach of war he also found himself helping to manage the application of radar to military purposes, his powers of organization as useful to government as they had been to Rutherford. From there it was a short step to involvement in the release of atomic energy. He did not go to Los Alamos, where the first bombs were made, but to Canada, to run a British–French–Canadian reactor project that carried the hopes of all three countries for postwar nuclear science. When the war ended he returned to England to become director of the Atomic Energy Research Establishment at Harwell, the post he held, as Sir John Cockcroft, when the Nobel telegram reached him in 1951.

Walton followed a quieter course. The mid-1930s was a period of diaspora for the Cavendish – Chadwick went to Liverpool, Blackett and Ellis to London, Oliphant to Birmingham, Mott to Bristol – and in 1934 Walton accepted a position at Trinity College in Dublin. It took a handsome gesture from his old university, which set aside its usual rigmarole of examinations and interviews to offer him a Fellowship purely on the basis of his academic distinction. Professional success may also have emboldened him in private life, for by this time he had proposed to Freda. He plucked up his courage during one of their meetings in Belfast as the couple strolled on Cave Hill, a beauty spot overlooking the sea, and she accepted him immediately. As soon as he was back in Ireland for good, therefore, and even before he had started his new job, Ernest and Freda were married in Dublin in a service conducted jointly by their fathers. Rutherford, long accustomed to weddings among his 'boys', sent the present he always sent: a copper tray. After a honeymoon in Switzerland Walton settled into a busy, happy life in Dublin in which a heavy teaching load left him almost no time for research. When war came Ireland was neutral and he remained where he was, although three times he received letters inviting him to join the British effort. The first was from C. P. Snow, by then a senior civil servant, and although Snow did not name a project it appears to have been an attempt to recruit him to radar work. The other two appeals came from Chadwick and Oliphant and though again secrecy prevented them identifying the job in question we know it was the atomic bomb. In each case Walton's reply was the same: most of the physics staff at Trinity had already left to support the British war effort and, with both his university and his country in desperate straits, he could not be spared.

Had he said yes he would have found himself working with Oliphant and Ernest Lawrence in California on electromagnetic techniques for refining the fissile uranium for the Hiroshima weapon. He never regretted missing this experience although he had no moral objection either to war work or to the atomic bomb. When the war ended he was quick to congratulate Oliphant on the 'amazing success' of his efforts. 'I had made up my mind it would be too big an effort to develop nuclear energy,' he wrote.[5] By 1951 he was professor at Trinity, a post he held until 1974.

Why did they have to wait nearly twenty years for the prize? It might have come earlier, for we know that Rutherford nominated them in 1937, but as the Nobel committee's minutes remain confidential we do not know the reasoning then or in later years. A glance at some of the other awards made in the 1940s and 1950s, however, removes much of the mystery. The physics winner in 1945, for example, was Wolfgang Pauli, for his formulation of the exclusion principle twenty-one years earlier. The winner in 1948 was Blackett, largely for his cloud-chamber work in the early 1920s. And as late as 1954 Max Born shared the prize for his role in developing quantum mechanics in the mid-1920s – a delay of almost thirty years. The strong impression is left that so much had been achieved in the study of the atom in the 1920s and 1930s that the Nobel committee had its hands full catching up, especially as it was also fulfilling its responsibility to recognize work in other branches of physics. Among others in the atomic field who received the prize ahead of Cockcroft and Walton were Heisenberg, Schrödinger, Dirac, Chadwick, Carl Anderson and, notably, Ernest Lawrence, who was honoured in 1939 (without Livingston) both for the invention of the cyclotron and for results obtained with it up to that year.

Modern perceptions of what was achieved in the Cavendish laboratory in 1932 are inevitably coloured by the atomic bombs and the Cold War that followed. The claim can be made that in the speculative excitement that followed the disintegration of lithium – that fuss that so dismayed Rutherford, Cockcroft and Walton – it was the press and the public who read the implications correctly, rather than the scientists. The newspapers spoke then of science soon achieving mastery over the vast potential of atomic energy, and just thirteen years later that potential was realized in the explosions over Japan. The scientists can be seen to have been over-cautious, or worse. Long before, H. G. Wells had compared humanity to 'a sleeper who handles matches in his sleep and wakes to find himself ablaze';[6] Rutherford, Chadwick, Cockcroft, Walton and others involved in nuclear science in the 1930s have sometimes been cast as the strikers of the match, the careless openers of a Pandora's box of troubles that would imperil humanity for ever. It would have been better, by implication, if they had left well alone.

This is a misjudgement. The atomic bomb and nuclear energy were familiar concepts in the Cavendish of 1932. Such notions could be traced back at least to the beginning of the century when the process of radioactive decay was first being unravelled and many scientists, Rutherford among them, pointed out that if nuclear disintegration could be brought under control it would offer an energy source of extraordinary power. Frederick Soddy, one of Rutherford's early collaborators, wrote in 1912 in a book entitled *The Interpretation of Radium*:

It can scarcely be doubted that one day we shall come to break down and build up elements in the laboratory as we now break down and build up compounds, and the pulses of the world will then throb with a new source of strength as immeasurably removed from any we at present control as they in turn are from the natural resources of the human savage.[7]

If this came about, Soddy suggested, the world might become a 'smiling Eden' where deserts could be made green, the North and South Poles could be thawed and humankind 'would have little need to earn its bread by the sweat of its brow'.[8]

It was on reading these predictions that Wells was moved to write *The World Made Free*, a book which, despite its sunny title, painted a much darker picture of the nuclear future. Published in 1914, it suggested that the essential atomic discovery would occur in 1933 and that a nuclear war would follow in 1955 that would leave millions dead. It was only out of the ruins left by that war that Soddy's 'smiling Eden' could be brought into existence. Though commercially unsuccessful by Wells's high standards (its appearance was overshadowed by the other events of 1914), the novel had considerable influence and opened up a new field of fiction for others to develop, among them the playwrights behind *Wings Over Europe*. By the early 1930s, however, it was rare to find a scientist speculating in the sort of airy terms used by Soddy twenty years earlier. He himself published a new book in 1932, *The Interpretation of the Atom*, and although it included a postscript reporting the Cockcroft and Walton breakthrough it was conspicuously less bold about the prospects for atomic power. Rutherford, when he was asked about such matters in these years, was usually cautious or dismissive, depending upon his mood. Matters had moved on. In general, the more physicists learned about the atom in those years the cooler and more sober became their view of the possibility of power and bombs. Even those who believed that it would probably come one day, and Cockcroft was among them, remained oppressively conscious of the many leaps of science and technology that would be required. It was natural for them to try to damp down public excitement when it arose.

Did the Cambridge breakthroughs open the way to the bomb? Certainly there are strong lines of connection, of which the most direct links Chadwick's neutron and a discovery in Berlin in 1938 by Otto Hahn and Fritz Strassmann. Using a beam of neutrons produced by the same technique employed by Chadwick in 1932, Hahn and Strassmann bombarded uranium and found to their astonishment that nuclei of this heaviest of all elements could split clean in two, a process which released energy on a scale far, far greater than the splitting of lithium nuclei. This was given the name 'fission'. Soon afterwards other researchers showed that fission was accompanied by a release of stray neutrons and this raised for the first time the possibility of something more than a minute laboratory-scale effect: a sustained chain reaction. Cockcroft and Walton had seen nuclei split in two and release energy, but those had been discrete and individual events on a tiny scale, so they were harmless. In a chain reaction a neutron fired at a uranium nucleus would cause it to split, releasing both energy and other neutrons. Those neutrons would immediately strike other, neighbouring uranium nuclei and the effect would be repeated. In a sufficient mass of uranium atoms a process of multiplication would follow so that one energy release would soon become one hundred, then ten thousand and then, within a fraction of a second, millions and billions and ultimately billions of billions, until the accumulated effect was an explosion on a catastrophic scale. This was a dire prospect, but no sooner had it loomed than theory stepped in to dispel it. Niels Bohr showed that ordinary uranium was incapable of sustaining a chain reaction; the effect would occur only in the variant known as uranium 235, and not only was this variant exceptionally rare but no means existed to separate it from ordinary uranium.

As late as 1939, then, the prospect of a nuclear bomb was still a distant one in the minds of most physicists, even if it was rather less so than it had been in 1932. It took the extreme circumstances of the first total war, a further flood of scientific and technical advances and above all the expenditure of the previously unimaginable sum of $2bn to make the bomb a reality. Accelerators would play their part, not only by supplying essential information about the nucleus but also by helping make possible the discovery of plutonium, the man-made radioactive element which was an alternative fissile material to uranium 235 and which provided the core for the Nagasaki bomb. Chadwick, Cockcroft, Walton and behind them Rutherford can all be said, therefore, to have contributed in important ways to the flow of discoveries that made possible the atomic bomb, but the sequence of events shows that it was not a simple matter of cause and effect. And just as it would be an exercise in anachronism to suggest that they should have foreseen this sequence so it would be wrong-headed to say that, knowing the nucleus might be dangerous, they should have made the decision to leave it alone.

As the citation made clear, when the Nobel committee honoured Cockcroft and Walton in 1951 it was not only for the fundamental character of their work but also for the legacy it left, and they did not have bombs and power stations in mind. With Lawrence, Livingston, Tuve, Lauritsen, Brasch, Lange and others they were the fathers of particle physics, a discipline which by 1951 had established itself internationally and would go on to grow beyond anything that even H. G. Wells might have conceived. By the time Cockcroft died in 1967 there were probably 1,500 particle physicists engaged in fundamental research around the world. Walton lived to 1995, when the number had passed 10,000.

Today the annual global budget of particle physics research is about £1bn and the accelerators have become so vast that in one or two cases they may be seen from space. At CERN in Geneva, at Brookhaven National Laboratory in New York State, at Fermilab in Illinois, at DESY in Hamburg and at a dozen other locations international teams of scientists plan and carry out experiments that push particles close to the speed of light, collide them with other particles and offer an ever-growing understanding of matter and the universe. To the four particles known in 1932 – proton, electron, neutron and positron – have been added neutrinos, pions, muons, quarks, gluons and many more. Problems that baffled Rutherford have long ago been resolved. There are, for example, two forces at work inside the nucleus, the strong and the weak, which are distinct from the familiar gravity and electromagnetic force. And electrons do not, after all, exist inside nuclei, or at least not in the normal sense: it appears that 'virtual' electrons may be present there (and everywhere else), becoming real only when conditions are right and in accordance with the quantum mechanical probabilities. The great modern quest is for an entity called the Higgs boson, and to help trap it a proposal has been made for an accelerator 33 kilometres in length, costing £4bn. But particle physics is by no means limited to fundamental research of this kind, indeed only 1 per cent of all the particle accelerators in the world are employed in such work. Of the remaining 10,000 half are used in medicine – for radiotherapy, biomedical research, the production of tracers and other purposes – and the other half belong in industry, where they produce new materials and alter existing ones, from car tyres and telephone wires to compact discs and semiconductors. Many of these machines still employ a Cockcroft–Walton apparatus as their first stage.

None of this, however, was in the minds of those two industrious and sober men in the spring of 1932, any more than bombs or power stations. What drove them was a force as fundamental to science as gravity and electromagnetism, a force personified in the figure of Ernest Rutherford, who inspired and supported them: curiosity. In June 1932, relenting a little from his ban on the press, Rutherford allowed a reporter from the *Daily Herald* into the Cavendish. After first delivering a lecture on the 'rot' and 'drivel'[9] usually to be found in the newspapers – he fished some crumpled cuttings from his pocket to illustrate the point – the laboratory director ushered his guest into the high-voltage room, which did not fail to impress. It was, wrote the young Ritchie Calder, 'a darkened chamber full of strange objects, rather weird but very wonderful, such as exists nowhere else in the world, not even in the imagination of Hollywood'.[10] He noted with awe the glowing glass tower of the rectifiers, the livid glare of lightning at the spark gap and the 'cannon' which fired protons at 6,000 miles per second ('across America in a second!'). He heard the hiss of the Burch pumps and, inside the humble observer's hut, witnessed the twinkle of scintillations on a zinc sulphide screen. And he marvelled at the way particles could be made to count themselves by means of wireless valves – Wynn-Williams had been at work. This was, he reported, 'the most scientific laboratory in the world' and for all the talk about the secrets of matter and the possibility of atomic energy, the physicists there loved their research 'not for its results, but for itself'. The last word he left to Lord Rutherford, who with a laugh offered his own simple justification for the taming of the atomic nucleus: 'We are rather like children, who must take a watch to pieces to see how it works.'

Postscript

Rutherford, having enjoyed a second youth experimenting with artificial particle beams, died in 1937, the victim of complications arising from what even then was a routine hernia operation. His death at the age of sixty-six provoked shock and grief among colleagues, students and friends, a tearful Bohr breaking the news to leading physicists at an international conference in Italy. Rutherford's ashes are buried in Westminster Abbey, close to Newton's grave. Those same colleagues, students and friends were thus left to cope without him as the Second World War transformed the discipline he did so much to create.

Several played prominent roles. Chadwick, who had won the Nobel prize for physics in 1935, led the British team at the Manhattan Project and Oliphant became Ernest Lawrence's deputy at a branch of the project in California, with Allibone in his group. Feather remained in Cambridge doing weapon-related nuclear research while Blackett and Mott spent much of the war developing operational research techniques and Eryl Wynn-Williams built machines for the codebreakers at Bletchley Park. Blackett, as we have seen, won the Nobel prize in 1948 and Mott followed much later in 1977. Chadwick ended his career as Master of his Cambridge college, Gonville and Caius, while Oliphant eventually became Governor of South Australia. Gamow, barred from leaving the Soviet Union in 1931, tried but failed to paddle his way from the Crimea to Turkey in a canoe. Allowed out on a short visa in 1933 to attend the Solvay conference he chose never to return and settled in the United States. There he found nuclear physics too crowded for his taste and switched first to cosmology and later to DNA theory, making important contributions to both. He enjoyed success as a popular writer, with books such as *Mr Tompkins Explores the Atom* and *One, Two, Three . . . Infinity*, and never lost his taste for mischief: perhaps his most famous stunt was to persuade Hans Bethe to co-sign a paper he had written with Ralph Alpher, so creating the Alpher-Bethe-Gamow theory in cosmology. He died in 1968. Peter Kapitza paid a visit to the Soviet Union in 1934 and was not allowed to leave, upon which Rutherford agreed to have most of his apparatus packed up (by Cockcroft) and shipped to Moscow. Kapitza endured difficult years under Stalin, including a long period effectively under arrest, but he went on to win a Nobel prize in 1978. Among the accelerator pioneers, Ernest Lawrence soon saw his cyclotron dominating the field in disintegration work, as he had known it would, and went on to become one of the statesmen of American science. Stanley Livingston, who felt his contribution at Berkeley had been overshadowed, moved east and built successful accelerators at Brookhaven. Tuve and Lauritsen enjoyed long and fruitful careers at their respective laboratories while Brasch and Lange fled Nazi Germany in opposite directions, Brasch going to the United States, where he pursued high-voltage work at Caltech, and Lange choosing the Soviet Union and eventually joining the Soviet nuclear weapons project. The continental theorists were also scattered by the rise of Nazism, Einstein choosing the United States, Schrödinger Ireland and Max Born Scotland. (Among Born's more unexpected legacies is his granddaughter, Olivia Newton-John.) Niels Bohr fled occupied Copenhagen, reluctantly, in 1943, spending time in Britain and the United States and failing to persuade Allied leaders of the danger of a nuclear arms race. He returned home as soon as peace came. Heisenberg remained in Germany through the war and contributed to the unsuccessful German nuclear project, as did Walther Bothe.

Freda and Ernest Walton enjoyed a long and happy marriage. In 1936 they too lost their first child, a girl named after her mother who died before she was a week old, but they went on to have another four children, all of whom grew up to pursue careers in science, three of them in physics. For her part Freda was always proud in later years to say she was among the first to know the atom had been split. She died in 1983, twelve years before Ernest.

As for the Cockcrofts, the baby that Elizabeth was expecting in that spring of 1932 was born in October and named Dorothea, and in the ensuing years the house on Sedley Taylor Road had to be extended to accommodate two more daughters. During the family's war travels these three were joined by a fourth daughter and then by a son. In age order the children became a nurse, a scientist, a teacher, a priest and an engineer. Though for periods they lived elsewhere, notably in Canada and at Harwell, the Cockcrofts always regarded Cambridge as their home and it was there that John died, suddenly, of a heart attack, in 1967. Elizabeth outlived her husband by two decades, dying in 1989 at the age of ninety. They are buried together in the same grave as Timothy.

Notes

Prelude: Manchester, 1909

1. Birks, *Rutherford at Manchester*, p. 8.
2. Marsden, Rutherford Memorial Lecture, *Proceedings of the Royal Society*, A226, 1954, pp. 296–7.
3. *Background to Modern Science*, ed. Needham and Pagel, p. 68. By Rutherford's account it was Marsden's supervisor, Geiger, who brought the news, but I have preferred Marsden's account, which contains detail and is more personal.
4. Ibid.
5. Eve, *Rutherford*, p. 198.
6. The phrase comes from Rowland, *Understanding the Atom*, p. 56. The author attributes the 'well-known comparison' to Lodge. Rutherford spoke of 'a gnat in the Albert Hall' (Wilson, *Rutherford, Simple Genius*, p. 573).

1. Cavendish

Walton's 19 October letter to his father and his file at the 1851 Archive were very helpful here, while for the character and role of the laboratory I have relied most on Crowther's history, Chadwick's American Institute of Physics interview and Hughes's *Radioactivists*.

1. Gray, *Cambridge*, p. 184.
2. Walton papers, Trinity College, Dublin, file 62, letter of 19/10/27.
3. Translation in Crowther, *The Cavendish Laboratory*, p. 49.

4. Walton papers, TCD, letter of 19/10/27.

5. Ibid.

6. Oliphant, *Rutherford: Recollections of the Cambridge Days*, p. 19.

7. Chadwick interview with Charles Weiner for the Center for History of Physics at the American Institute of Physics, April 1969, p. 41. Subsequent quotations are identified as Chadwick/Weiner.

8. Walton papers, TCD, file 1, Methodist College school report, Christmas 1920.

9. Walton file, Archive of the 1851 Royal Commission, letter of 15/7/27.

10. Walton papers, TCD, file 52, 4/10/27.

2. *'Mollycewels an' Atoms'*

The portrait of Rutherford is gleaned from the large literature of tributes and biographies, leaning most heavily on Oliphant and Wilson. The account of the atom draws on various popular books and textbooks of the time and on Hendry. For Rutherford's difficulties in the later 1920s Chadwick's AIP interview is again useful, while Feather's biography of Rutherford and his AIP interview also shed valuable light.

1. Wilson, *Rutherford*, p. 90.

2. Chadwick/Weiner, p. 122.

3. Marsden's Rutherford Memorial Lecture, p. 294.

4. Obituary of Crowe in *Nature*, vol. 211, 1966, p. 20.

5. Interview with William Kay, Blackett papers, Royal Society, file H80.

6. Ibid.

7. Oliphant, *Recollections*, p. 120.

8. Eve, *Rutherford*, p. 342.

9. Wilson, *Rutherford*, p. 275.

10. Allibone interview with the author, March 1998.

11. O'Casey, S., 1926, *The Plough and the Stars* (London, Macmillan), p. 15.

12. The modern term is 'relative atomic mass'.

13. Chadwick/Weiner, p. 70.

14. Andrade, *Structure*, p. 140.

15. Chadwick/Weiner, p. 70.

3. Method

Hughes, Brown and the Chadwick AIP interview tell the story of the Vienna controversy, while Blackett is one of several to testify to the miseries of particle counting in this period. The original published papers, and particularly the Chadwick paper that is quoted, give descriptions of the techniques used.

1. Eve, *Rutherford*, p. 299.

2. Blackett papers, Royal Society, file B116, 'Big Machines in Physics', 1954.

3. *Philosophical Magazine*, vol. 2, November 1926, pp. 1056–75.

4. Brown, *Neutron*, p. 87.

5. Ibid.

6. Chadwick/Weiner, p. 62.

7. Ibid.

8. See Heilbron and Seidel, *Lawrence*, p. 48n.

9. D. C. Rose, quoted in Hughes, *Radioactivists*, p. 131.

4. A Way Forward

My interview with Professor Allibone was invaluable for the second half of the chapter, providing descriptions of the room and the people and filling in his own biography. I also made use of his biography of Cockcroft (written with Hartcup) and of various Cockcroft papers. Walton's early experiments are best described in his reports to the 1851 Commission and the published paper.

1. The speech appears in *Proceedings of the Royal Society*, A117, 1928, pp. 300–316.
2. Chadwick/Weiner, p. 52.
3. Allibone interview with author.
4. Hartcup and Allibone, *Cockcroft*, p. 13.
5. Undated transcription in the Cockcroft family's private collection of letters.
6. Public Record Office, DSIR 3/42, Rutherford letter of 1/6/25.
7. Hendry, *Cambridge Physics*, pp. 157–8.

5. *A Man in White Trousers*

Gamow's pre-1933 personal papers were lost when he fled the Soviet Union. This chapter relies mainly on his autobiography, his AIP interview, correspondence in the Archive for the History of Quantum Physics and Stuewer. Helpful introductions to quantum mechanics can be found in Eddington and in Gamow's *Thirty Years That Shook Physics*, although I was also grateful for guidance from Andrew Whitaker and Alan Walton.

1. Gamow, *My World Line*, p. 15.
2. Gamow interview with Charles Weiner, April 1968, for the Center for History of Physics at the American Institute of Physics, p. 3.
3. Gamow, *My World Line*, p. 51.
4. Joffe to Bohr, undated, AHQP, microfilm E4, BSC12.
5. Gamow, *My World Line*, p. 55.
6. Ibid.
7. Ibid.
8. Ibid., p. 64.

6. A Finite Probability

Again the AHQP papers contributed a good deal, in particular when it came to fixing the chronology, while Casimir is especially good on the mood in Copenhagen.

 1. Mott, *A Life in Science*, pp. 25–6.
 2. Ibid., p. 29.
 3. Casimir, *Haphazard Reality*, p. 118.
 4. Mott, *A Life*, p. 28.
 5. *Nature*, vol. 112, 1928, p. 439.
 6. Fowler to Bohr, 5/10/28, AHQP, E4, BSC10.
 7. Cockcroft interview with Thomas S. Kuhn, May 1963, for the Center for History of Physics at the American Institute of Physics, p. 11.
 8. Cockcroft papers, Churchill Archives Centre, Churchill College, Cambridge (henceforth CAC), 20/80.
 9. Eve, *Rutherford*, p. 304.
10. De Bruyne, *My Life*, p. 58.
11. Ibid., pp. 58–9.
12. *Proceedings of the Institution of Mechanical Engineers*, vol. 2, 1928, pp. 623–4.
13. Ibid.
14. Mott to Bohr, February 1929, AHQP, E4, BSC14.
15. *Proceedings of the Royal Society*, A123, 1929, pp. 387–8.
16. Letter to E. M. McMillan, 1/9/1977, in Stuewer, *Nuclear Physics*, p. 147.

7. Hardware

The story of the construction of the two accelerators, told in this and later chapters, was assembled mainly from Walton's reports to the 1851 Commission, his Ph.D. thesis and the Cockcroft-Walton published papers of 1930 and 1932. As the references show, some additional details emerged from Walton's correspondence with Freda Wilson and from Cockcroft's with his friends at Metro-Vick, in particular George McKerrow, while there is some information in their later Nobel addresses. Allibone, too, is an important witness – it was he, for example, who alerted Cockcroft to the Burch pumps.

1. Hendry, *Cambridge Physics*, p. 160.
2. Cockcroft papers, CAC, 20/59, letter of 6/8/29.
3. The words were spoken to J. K. Roberts and recalled by Edward Bullard. They appear in the entry on Harrie Massey in *Biographical Memoirs of Fellows of the Royal Society*, 1984, p. 482.

8. Lab Life

Oliphant, Feather's writings, Hendry and the Rutherford Memorial Lectures of the 1940s and 1950s provide the backbone of this chapter, while as the references indicate other crumbs about the lab have been plucked from across the literature.

1. Chadwick/Weiner, p. 54.
2. H. A. Bumstead to J. J. Thomson, quoted in Thomson, *Recollections*, pp. 143–4.
3. Cockcroft papers, CAC, 20/59, letter of 3/12/1929.
4. From 'Reminiscences of the Cavendish Laboratory', the typescript of a talk by Norman Feather, in the Chadwick papers at the Churchill Archives Centre, II 2/1.

5. Chadwick/Weiner, p. 56.
6. Cockcroft papers, CAC, letter of 22/10/1924.
7. Walton papers, TCD, file 60, Ernest Walton to Winifred Wilson, 13/11/32. Further quotations from this correspondence, which is all in file 60, appear simply as 'EW to WW' or 'WW to EW'.
8. Peierls, *Bird of Passage*, pp. 35–6.
9. Boag *et al.*, *Kapitza*, p. 139.
10. EW to WW, 9/11/31.
11. Casimir, *Haphazard Reality*, p. 74.
12. Cockcroft to Elizabeth Crabtree, 18/5/24, from the Cockcroft family's collection of papers.
13. Cockcroft to E. Crabtree, August 1925, family collection.
14. Hartcup and Allibone, *Cockcroft*, p. 35.
15. Cockcroft, letter of 1925, family collection.
16. Hartcup and Allibone, *Cockcroft*, p. 43.
17. Letter of the Crabtrees to Elizabeth Cockcroft, 8/11/1929, in the family collection.
18. Ibid.
19. Hartcup and Allibone, *Cockcroft*, p. 44.
20. Walton file, 1851 Archive, letter of 22/5/1929.
21. Walton interview with author, 1990.
22. Chadwick/Weiner, p. 47.

9. *Other Ideas*

The files in the 1851 Archive on Wynn-Williams and Harold Cave appear to be the best sources on the early Cavendish counter work; the material in Hendry dwells mainly on slightly later developments. Brown, Hughes and Chadwick's AIP interview tell the polonium story. On Tuve I have relied largely on Cornell and the AIP interview, while the monthly Department of Terrestrial Magnetism reports offer a detailed chronology. Burghard Weiss kindly helped me with the German story.

1. Sargent, *Physics in Canada*, vol. 36, 5, 1980, p. 97.
2. Chadwick/Weiner, p. 1.
3. Ibid., p. 18.
4. Ibid., p. 30.
5. Ibid., p. 49.
6. Merle Tuve interview with Charles Weiner, March 1967, for the Center for the History of Physics at the American Institute of Physics, p. 12.
7. Monthly reports of the Department of Terrestrial Magnetism, Carnegie Institution, 13/1/1929. Copies of these reports are held at the Niels Bohr Library of the American Institute of Physics.
8. Stuewer, *Nuclear Physics*, p. 135.
9. Charles Lauritsen interview with Charles Weiner, June 1966, for the Center for the History of Physics at the American Institute of Physics, p. 1.
10. Davis, *Lawrence and Oppenheimer*, p. 28.

10. Turning Point

The same sources used in Chapter 7 tell the story of the apparatus, although my account of the rationale for the switch to a larger machine is largely a matter of deduction, neither man having explained it at any length. For the personal content the collective memories of the Walton and Cockcroft families helpfully supplemented the written record.

1. Wright (ed.), *University Studies*, p. 70.
2. Ibid.
3. Walton papers, TCD, file 4, a text of Cockcroft's Nobel lecture. The passage including these remarks was omitted from the final version, perhaps to avoid overlapping with Walton's lecture.

4. Public Record Office, DSIR 2/386, 13/6/1930.
5. Laurence file, 1851 Archive, letter of 29/11/1929. I am grateful to his daughter, Patricia Buchanan, for permission to quote from this.
6. Ibid.
7. *Proceedings of the Royal Society*, A129, 1930, pp. 477–89.
8. EW to WW, 29/7/1930.
9. WW to EW, 12/8/1930.
10. WW to EW, 18/9/1930.
11. EW to WW, 22/2/1931.
12. EW to WW, 17/5/1931.
13. EW to WW, 18/1/1931.
14. Rutherford papers, Cambridge University Library, PA 138.
15. Heilbron and Seidel, *Lawrence*, p. 88.

11. *Off to the Races*

Titbits about Gamow's activities in this period turn up in many memoirs, while Crowther's papers in Brighton are also useful. Feather's AIP interview tells the story of his American trip and is complemented by Chadwick's. I am grateful to Alf Refsum at Queen's University, Belfast for help on the multiplier circuit. At this point I must mention the name of William Birtwhistle, a Cavendish technician who would later become a mainstay of the high-voltage lab and who may have helped with this early construction work. Birtwhistle was hired by Cockcroft in 1930 but whether he was directly involved with the accelerator before the spring of 1932 I have been unable to establish with any confidence, the scraps of evidence I have found tending to contradict one another. A letter from Cockcroft to the DSIR of 14 June 1932 (CAC papers, 20/28) speaks only of him helping 'during the last two months'.

1. Carnegie Institution, DTM monthly reports, Breit, 16/1/1929.
2. Rockefeller Archive Center, New York, Gamow file, assessor's recommendation of 28/6/1929, p. 2.
3. Moore, *Bohr*, pp. 145–6.
4. Casimir, *Haphazard Reality*, p. 117.
5. Williamson (ed.), *Making of Physicists*, p. 41.
6. Gamow, *Constitution*, p. 1.
7. Bohr to Rutherford, 3/2/1930, AHQP, E4, BSC25.
8. Brown, *Neutron*, p. 102.
9. Ibid., p. 169.
10. Norman Feather interview with Charles Weiner, February 1971, for the Center for the History of Physics at the American Institute for Physics, p. 33.
11. Ibid., p. 35.
12. Cockcroft papers, CAC, 20/10A, letter of 15/10/1930.
13. EW to WW, 2/12/1930.
14. Cockcroft papers, CAC, 20/28, 28/1/1931.
15. Condensers are known today as capacitors.
16. EW to WW, 3/3/1931.
17. WW to EW, 1/3/1931.
18. EW to WW, 3/3/1931.
19. EW to WW, 6/5/1931.
20. EW to WW, 31/5/1931.
21. Stanley Livingston interview with Charles Weiner and Neil Goldman, August 1967, for the Center for History of Physics at the American Institute of Physics, p. 11.
22. Heilbron and Seidel, *Lawrence*, p. 93.
23. Davis, *Lawrence and Oppenheimer*, p. 33.
24. Livingston interview with Weiner, pp. 17–18.
25. Davis, *Lawrence and Oppenheimer*, p. 37.
26. Childs, *American Genius*, p. 164.
27. Heilbron and Seidel, *Lawrence*, p. 99.
28. Ibid., p. 100.

29. Davis, *Lawrence and Oppenheimer*, p. 40.
30. Tuve interview with Weiner, p. 16, edited from Tuve's account. Townsend is J. S. E. Townsend, the Oxford professor of physics.
31. Carnegie Institution, DTM monthly reports, September 1931.

12. *Timeliness and Promise*

On Chadwick and the neutron the original papers are extremely clear. It is worth pointing out that at this stage Chadwick still thought of the neutron as an intimate combination of proton and electron, a view no longer taken.

1. Hartcup and Allibone, *Cockcroft*, p. 48.
2. Ibid., p. 49.
3. *Proceedings of the Royal Society*, A136, 1932, p. 625.
4. Ibid.
5. EW to WW, 19/11/1931.
6. EW to WW, 29/11/1931.
7. Ibid.
8. Hendry, *Cambridge Physics*, p. 45.
9. Feather interview with Weiner, p. 43.
10. Hendry, *Cambridge Physics*, p. 45.
11. Chadwick/Weiner, p. 72.
12. 'The Existence of a Neutron', *Proceedings of the Royal Society*, A136, 1932, p. 696.
13. Snow, *Physicists*, p. 85.
14. 'Existence of a Neutron', p. 697.
15. Ibid.
16. Oliphant, *Recollections*, p. 77.
17. Snow, *Physicists*, p. 85.
18. *Physical Review*, vol. 38, 1931, p. 1919.
19. *Physical Review*, vol. 39, 1932, p. 384.
20. Ibid., p. 834.

21. This and the subsequent quotations from Boyce's letter are taken from Charles Weiner's '1932 – Moving into the New Physics', *Physics Today*, May 1972, pp. 40–42.
22. *Nature*, vol. 129, 1932, p. 242.
23. WW to EW, 21/1/1932.
24. EW to WW, 25/1/1932.
25. EW to WW, 13/3/1932.
26. Ibid.
27. Public Record Office, DSIR 1/17.
28. *Physical Review*, vol. 40, 1932, pp. 19–36.

13. Red Letter Day

Walton recounted his part of this story many times and the account he gave in *Europhysics News* is among the fullest, but even for that he does not seem to have consulted his letter to Freda of 17/4, which provides a good deal of the additional detail in this chapter. Walton's laboratory notes, held at the Churchill Archives Centre, are also helpful.

1. Lecture by Lord Bowden, 15/3/1979, pp. 18–19.
2. Chadwick papers, CAC, II 2/1, pp. 18–19.
3. Cockcroft papers, CAC, 20/28, letter to Karl Darrow, 1/2/1938.
4. Hartcup and Allibone, *Cockcroft*, p. 50.
5. Ibid.
6. Oliphant, *Recollections*, p. 85. Twenty-five years later (Walton papers, file 56, 12/12/1957) Walton suggested this was 600,000 or 700,000 volts but after that he used to decline to give a figure, implying that he had had second thoughts. In the succeeding weeks they seem not to have exceeded 550,000 volts.
7. Ibid., p. 86.

8. Ibid.

9. Snow, *Physicists*, p. 89.

10. Hartcup and Allibone, *Cockcroft*, p. 52.

11. *Nature*, vol. 129, 1932, p. 649.

12. Ibid.

13. EW to WW, 17/4/1932. This letter, from which the quotations beneath are also drawn, usefully clears up a long-standing muddle about whether the breakthrough took place on 13 or 14 April. Written on Sunday 17, it not only explicitly identifies the previous Thursday (14th) but also plots clearly the events of the days in between.

14. WW to EW, 20/4/1932.

15. Ibid.

16. Hendry, *Cambridge Physics*, p. 99.

17. This letter is quoted in full in Bohr, N., 1963, *Essays 1958–1962 on Atomic Physics and Human Knowledge* (London, Clay, 1963), pp. 67–8.

18. Eve, *Rutherford*, pp. 356–8.

19. The text of Rutherford's talk, from which these and the subsequent quotations are taken, can be found in *Proceedings of the Royal Society*, A136, 1932, pp. 735–62.

20. EW to WW, 1/5/1932.

21. Allibone, in Hartcup and Allibone, *Cockcroft*, p. 53. Some of the additional disintegration effects mentioned by Rutherford that night would prove to be the results of contamination in the acceleration tube.

14. Still Safe

The Crowther papers were again helpful here and Crowther's long article on the breakthrough for *The Nineteenth Century*, published that July, is also well worth reading.

1. This, and the remarks on Bohr and Dirac, can be found in Crowther papers, University of Sussex, file 32, notes on his Danish visit.
2. Evans, *Man of Power*, p. 185.
3. EW to WW. The letter is dated 1/5/1932 but this was in a PS added before posting on the 2nd.
4. *Daily Mail*, 2/5/1932.
5. *Daily Mirror*, 3/5/1932.
6. Nichols and Browne, *Wings Over Europe*, p. 90.
7. Ibid., p. vii.
8. EW to WW, 8/5/1932.
9. WW to EW, 4/5/1932.
10. WW to EW, 10/5/1932.
11. EW to FW, 8/5/1932.
12. EW to WW, 8/5/1932.
13. Letter of 11/5/1932, family collection.
14. Letter from Uncle Jos, May 1932, family collection.
15. Letter from John's mother, early May 1932, family collection.
16. EW to WW, 8/5/1932.
17. Einstein letter to Moritz Schlick from Oxford, Albert Einstein Archives 21–640, 18/5/1932 ('*Staunen und Bewunderung*' in the original). Quoted by permission of the Albert Einstein Archives, Hebrew University of Jerusalem.
18. Crowther papers, box 132, 3/5/1932.
19. Eve, *Rutherford*, p. 360.
20. *New York Times*, 8/5/1932, section 9, p. XX.
21. Childs, *American Genius*, p. 181.
22. Ibid.
23. Davis, *Lawrence and Oppenheimer*, p. 45.
24. Livingston interview with Weiner, p. 25.
25. Merle Tuve interview with Thomas D. Cornell, January–February 1982, Center for History of Physics, American Institute of Physics, p. 44.

26. Cockcroft papers, CAC, 20/28, letter to Darrow, 1/2/1938.
27. Goldsmith, *Frédéric Joliot-Curie*, p. 48.

15. Nobel

1. Walton papers, TCD, file 4, 15/11/1951.
2. Ibid., file 53, 19/11/1951.
3. *Nobel Lectures: Physics 1942–62*, p. 165.
4. *Les prix Nobel en 1951*, p. 66.
5. Walton papers, TCD, file 55, letter of 8/11/1945.
6. Wells, *The World Set Free*, p. 95.
7. Soddy, *Interpretation of Radium*, p. 238.
8. Ibid.
9. Wilson, *Rutherford*, p. 572.
10. *Daily Herald*, 'The Truth about the Atom', 27/6/1932.

Acknowledgements

The story told in these pages may concern a landmark event in science but it has not always had the attention it deserves. Rutherford's numerous biographers have tended to devote more pages to the heroics of his youth, when he led the scientific charge personally. Cockcroft and Walton, meanwhile, were not the sort to draw attention to their own success and, to make matters worse, by the time the award of the Nobel prize obliged them to give an account of their work their recollection had become hazy. Between them, in fact, they managed to muddle up several details, including the sequence of events at the beginning and the date of the climax. Among their Cavendish colleagues only Thomas Allibone was a true witness and it is no accident that he was Cockcroft's biographer – although even he was not present in the lab for the final phase. (I was fortunate to interview Professor Allibone in the course of my researches; I also interviewed Walton himself in 1990.) The principal development that made possible a fuller account was the opening up of Walton's personal archive at Trinity College, Dublin, following his death in 1995, for the many letters and papers, notably the correspondence with Winifred, gave much additional depth and perspective to the story. I am indebted to the Walton family, and particularly to Alan, for permission to quote from those papers and for a great deal of other support over the years. The staff at TCD library were most helpful. Cockcroft papers survive in abundance, mostly at the Churchill Archives Centre in Cambridge but also at the Public Record Office, and I have been made welcome at both institutions. I must also thank Christopher and Dorothea Cockcroft for allowing me access to the family's private collection of papers, for permission to quote and for their warm hospitality. Next I should pay tribute to the Center for History of Physics of the American Institute of Physics in College Park, Maryland, where Spencer Weart, the director, provided invaluable support and Katherine Hayes, associate archivist, was not only extremely helpful and patient but also most friendly and encouraging. From the Center's Niels Bohr Library I was able to borrow a number of the fine oral history interviews conducted mainly by Charles Weiner with the likes of Chadwick, Feather, Cockcroft, Gamow, Tuve and Livingston, which have been essential in building up the picture. The Center is also responsible for compiling the Archive for the History of Quantum Physics, including a large microfilm collection of scientists' correspondence which I was able to consult at Imperial College, London. Of the other historical collections I have consulted I should mention first the Archive of the Royal Commission for the 1851 Exhibition, where once again Valerie Phillips steered me ably through the files. Then there are the Blackett papers at the Royal Society, the Crowther papers at Sussex University, the Rutherford papers at Cambridge University Library and the Chadwick papers at the Churchill Centre. From overseas I have received help from Felicity Pors at the Niels Bohr Archive in Copenhagen, Barbara Wolff at the Albert Einstein Archives in Jerusalem and the staff of the Rockefeller Archive Center in New York. I am grateful to all, and I must pay special tribute to John F. Zeugner of Worcester Polytechnic Institute in Maryland, whose efforts on my behalf were unstinting. I should also thank the courteous, efficient staff of the British Library, where much of my research was conducted and which became a second home at times. A number of people read drafts or parts of drafts for me and offered many suggestions and corrections in the effort to keep me on the path of scientific sense, and to them I am especially grateful. They include, notably, Alan Walton and Christopher Cockcroft, physicist and engineer respectively. Bill Burcham and Alan Segar were rigorous, thorough and helpful critics on the physics front, as were Per F. Dahl, Andrew Whitaker and Burghard Weiss on the scientific history. Bill Burcham, a veteran of the Rutherford Cavendish, who knew the high-voltage room well, also supplied vivid insights about it and the life of the lab. Over the whole length of the project Tom Wilkie has been, as ever, a lucid and indulgent scientific guide. None of the above, of course, bears responsibility for any errors in the final text. Penny Jones, who was in at the start, was another supportive critic of the work in progress. Others I should thank for advice, help, encouragement or sustenance of various kinds over the years include Gail Gilliland and Keith Clarke, Rod Home, Gordon Squires, Stuart Cull-Candy, Ian Jack, Terry Volk, Ross Galbreath, Brum and Patricia Henderson, Tess Poole, Alf Refsum, Allen Packwood, Gerard O'Brien, Peter Roebuck, Lorna Arnold, Jacinta Kelly, Andrew Brown, Patricia Buchanan, Ann Robertson and, chief among the long sufferers, my family and my neighbours in Muswell Hill.

Bibliography

Of the works listed below a few were especially valuable. Wilson's biography of Rutherford is outstanding while Oliphant's recollections provide the most vivid single account of Cavendish life. Hendry is a fine anthology of reminiscences, well introduced, while the Hartcup and Allibone biography of Cockcroft and the Brown life of Chadwick were most useful. I also leant heavily on Cornell's life of Tuve, Stuewer on Gamow, Hughes's *Radioactivists* and the Standley, McAvoy and Mason work on Walton. The best general resource for biographical matters was the *Biographical Memoirs of Fellows of the Royal Society*. Fuller bibliographies in this field, embracing all the scientific papers, can be found in Heilbron and Seidel and in Beyer, while Lowood is a helpful guide to Rutherford's remarkable output of articles and lectures.

Books

Allibone, T. E., 1976, *The Royal Society and its Dining Clubs* (Oxford, Pergamon)

—, 1973, *Rutherford, the Father of Nuclear Energy* (Manchester University Press)

Andrade, E. N. da C., 1927, *The Structure of the Atom* (London, Bell)

Beyer, R. T., 1949, *Selected Papers in Foundations of Nuclear Physics* (New York, Dover)

Birks, J. (ed.), 1962, *Rutherford at Manchester* (London, Heywood)

Boag, J. W. et al. (eds.), 1990, *Kapitza in Cambridge and Moscow: Life and Letters of a Russian Physicist* (Amsterdam, North-Holland)

Boorse, H. A., *The Atomic Scientists: A Biographical History* (London, Wiley)

Brown, A., 1997, *The Neutron and the Bomb: A Biography of Sir James Chadwick* (Oxford University Press)

Bunge, M., and Shea, W. R. (eds.), 1979, *Rutherford and Physics at the Turn of the Century* (New York, Dawson)

Campbell, J., 2000, *Rutherford: Scientist Supreme* (Christchurch, ASS Publications)

Casimir, H. B. G., 1983, *Haphazard Reality: Half a Century of Science* (New York, Harper Colophon)

Childs, H., 1968, *An American Genius: The Life of Ernest Orlando Lawrence* (New York, Dutton)

Cockburn, S., and Ellyard, D., 1981, *Oliphant: The Life and Times of Sir Mark Oliphant* (Adelaide, Axiom)

Crowther, J. G., 1974, *The Cavendish Laboratory 1874–1974* (London, Macmillan)

Darrow, K. K., 1936, *The Renaissance of Physics* (New York, Macmillan)

Davis, N. P., 1986, *Lawrence and Oppenheimer* (New York, Da Capo)

de Bruyne, N., 1996, *My Life* (Cambridge, Midsummer)

Dummelow, J., 1949, *Metropolitan-Vickers Electrical Company 1899–1949* (Manchester)

Eddington, A. S., 1929, *The Nature of the Physical World* (Cambridge University Press)

Elsasser, W. M., 1978, *Memoirs of a Physicist in the Atomic Age* (Bristol, Hilger)

Evans, I. B. N., 1939, *Man of Power: The Life Story of Baron Rutherford of Nelson O.M., F.R.S.* (London, Stanley Paul)

Eve, A. S., 1939, *Rutherford, Being the Life and Letters of the Rt. Hon. Lord Rutherford, O.M.* (Cambridge University Press)

Feather, N., 1940, *Lord Rutherford* (London, Blackie)

—, 1936, *An Introduction to Nuclear Physics* (Cambridge University Press)

Gamow, G., 1993 edn, *Mr Tompkins in Paperback* (Cambridge, Canto)

——, 1970 edn, *My World Line: An Informal Autobiography* (New York, Viking)

——, 1966, *Thirty Years That Shook Physics* (New York, Dover)

——, 1931, *Constitution of Atomic Nuclei and Radioactivity* (Oxford University Press)

Geiger, H., and Makower, W., 1912, *Practical Measurements in Radio-Activity* (London, Longmans)

Goldsmith, M., 1976, *Frédéric Joliot-Curie: The Man and His Theories* (London, Lawrence & Wishart)

Gray, A., 1925, *The Town of Cambridge* (Cambridge, Heffers)

Hartcup, G., and Allibone, T. E., 1984, *Cockcroft and the Atom* (Bristol, Adam Hilger)

Heilbron, J. L., and Seidel, R. W., 1989, *Lawrence and His Laboratory* (University of California Press)

Hendry, J. (ed.), 1984, *Cambridge Physics in the Thirties* (Bristol, Adam Hilger)

Howarth, T. E. B., 1978, *Cambridge Between Two World Wars* (London, Collins)

Jeans, J., 1930, *The Universe Around Us* (Cambridge University Press)

Korsunsky, M., 1963, *The Atomic Nucleus* (New York, Dover)

Livanova, A., 1980, *Landau: A Great Physicist and Teacher* (Oxford, Pergamon)

Livingston, M. S., and Blewett, J. P., 1962, *Particle Accelerators* (New York, McGraw-Hill)

Lodge, O., 1927, *Modern Scientific Ideas* (London, Benn)

——, 1924, *Atoms and Rays* (London, Benn)

Lowood, H., 1979, *Ernest Rutherford: A Bibliography of His Non-technical Writings* (University of California)

Marsden, E., 1969, *Sir Ernest Marsden 80th Birthday Book: A Tribute from His Friends and Colleagues* (Wellington, Reed)

Moore, W., 1994, *A Life of Erwin Schrödinger* (Cambridge University Press)

—, 1967, *Niels Bohr: The Man and the Scientist* (London, Hodder)

Mott, N., 1986, *A Life in Science* (London, Taylor & Francis)

Nichols, R., and Browne, M., 1932, *Wings over Europe* (London, Chatto)

Nobel Institute, 1964, *Nobel Lectures: Physics 1942–1962* (Amsterdam, Elsevier)

—, 1952, *Les prix Nobel en 1951* (Stockholm)

Oliphant, M., 1972, *Rutherford: Recollections of the Cambridge Days* (Amsterdam, Elsevier)

Peierls, R., 1985, *Bird of Passage: Recollections of a Physicist* (Princeton University Press)

Robertsen, P., 1979, *The Early Years: The Niels Bohr Institute 1921–30* (Copenhagen, Akademisk Forlag)

Romer, A., 1960, *The Restless Atom: The Awakening of Nuclear Physics* (New York, Dover)

Rowland, J., 1938, *Understanding the Atom* (London, Gollancz)

Rutherford, E., 1954, *Rutherford by Those Who Knew Him* (London, Institute of Physics)

Rutherford, E., Chadwick, J., and Ellis, C. D., 1930, *Radiations from Radioactive Substances* (Cambridge University Press)

Snow, C. P., 1982, *The Physicists: A Generation That Changed the World* (London, Papermac)

Soddy, F., 1932, *The Interpretation of the Atom* (London, Murray)

—, 1912, *The Interpretation of Radium* (London, Murray)

Stuewer, R. H. (ed.), 1979, *Nuclear Physics in Retrospect* (University of Minnesota Press)

Thomson, J. J., 1936, *Recollections and Reflections* (London, Benn)

Waloschek, P., 1994, *The Infancy of Particle Accelerators: The Life and Work of Rolf Wideröe* (Online: www-library.desy.de/elbook.html)

Weisskopf, V., 1991, *The Joy of Insight: Passions of a Physicist* (New York, Basic)

Wells, H. G., 1914, *The World Set Free* (London, Macmillan)

Williamson, R., (ed.), 1987, *The Making of Physicists* (Bristol, Adam Hilger)

Wilson, D., 1983, *Rutherford, Simple Genius* (London, Hodder)

Wilson, E. J. N., 2001, *An Introduction to Particle Accelerators* (Oxford University Press)

Articles and theses

Blackett, P. M. S., 1933, 'The Craft of Experimental Physics', in Wright H. (ed.), *University Studies*, pp. 67–96 (London, Nicholson)

Bowden, B. V. (Lord), 1979, 'Professor the Lord Rutherford of Nelson: Christchurch's Most Famous Son', a lecture delivered on 15 March 1979 at Canterbury University in New Zealand

Burrill, E. A., 1967, 'Van de Graaff, the Man and His Accelerators', *Physics Today*, 20, 2, p. 47

Cathcart, B., 2003, 'Ernest Walton 1903–1995', in McCartney, M., and Whitaker, A. (eds.), *Physicists of Ireland: Passion and Precision* (Bristol, Institute of Physics)

—, 1992, 'Ernest Walton, Atomic Scientist', in O'Brien, G., and Roebuck P. (eds.), *Nine Ulster Lives* (Belfast, UHF)

'Cavendish Laboratory Supplement', 1926, *Nature*, 298, December

Cornell, T. D., 1986, 'Merle A. Tuve and His Program of Nuclear Studies at the Department of Terrestrial Magnetism: the Early Career of a Modern American Physicist', Ph.D. thesis, Johns Hopkins University, Baltimore, Md.

Crowther, J. G., 1932, 'Disintegrating Atoms by Machinery', *The Nineteenth Century*, July, pp. 1–13

Devons, S., 1971, 'Recollections of Rutherford and the Cavendish', *Physics Today*, December, pp. 39–45

Geiger, H., 1938, 'Memories of Rutherford in Manchester', *Nature*, 141, August, p. 244

Hughes, J. A., 1993, 'The Radioactivists: Community, Controversy and the Rise of Nuclear Physics', Ph.D. thesis, Corpus Christi College, Cambridge

Kapitza, P. L., 1966, 'Recollections of Lord Rutherford', *Proceedings of the Royal Society*, A294, pp. 123–37

Pierce, R. C., 1937, 'Electricity Supply in Cambridge', *The Cam*, April, pp. 108–9

Rutherford, E., 1938, 'Forty Years of Physics', in Needham J., and Pagel W. (eds.), *Background to Modern Science*, pp. 49–74 (Cambridge University Press)

Sargent, B. W., 1980, 'Recollections of the Cavendish Laboratory Directed by Rutherford', *Physics in Canada*, 36, 4, July, pp. 75–80 and 5, August, pp. 97–100

Standley, R. L., McAvoy, W., and Mason, P., 1989, 'The Life and Scientific Achievements of Dr Ernest Thomas Sinton Walton', degree thesis, Worcester Polytechnic Institute, Worcester, Md.

Stuewer, R. H., 1986, 'Gamow's Theory of Alpha-Decay', in Ullmann-Margalit, E. (ed.), *The Kaleidoscope of Science*, pp. 147–186 (Dordrecht, Reidel)

Trenn, T., 1986, 'The Geiger-Müller Counter of 1928', *Annals of Science*, 43, pp. 111–35

Walton, E. T. S., 1982, 'Recollections of Nuclear Physics in the Early Nineteen Thirties', *Europhysics News*, 13, August, pp. 1–3

Weiner, C., 1972, '1932 – Moving into the New Physics', *Physics Today*, 25, May, pp. 40–49

Weiss, B., 1999, 'Blitze für Kernphysik und Strahlentherapie . . .', *Technikgeschichte*, 66, 2

Index

accelerator 56, 59, 97, 99, 100, 111, 136, 149, 151, 154, 156, 159, 173, 175, 196, 200, 206, 207, 252, 260, 268, 269, 273
Adelaide 19
Albert, Prince 16–17
Allgemeine Elektrizitäts-Gesellschaft (AEG) 147, 148, 174
Allibone, Thomas 59, 60, 62, 63, 64, 96, 97, 103, 106, 107, 108, 119, 121, 122, 124, 127, 139, 153, 158, 173, 175, 182, 187, 190, 222, 240, 272
alpha escape theory 89, 93, 98
alpha particle 32, 37, 38, 39, 40, 42, 52, 53, 80, 81, 88, 90, 92, 93, 132, 134, 135, 140, 141, 143, 165, 180, 181, 183, 195, 209, 213, 219, 227, 228, 232, 233, 236, 237, 242, 256
alpha-ray 58
Alpher, Ralph 273
Alpher-Bethe-Gamow theory 273
aluminium 41, 181, 242
Amsterdam 201
Anderson, Carl 259, 260, 265
anode 108, 111, 153, 154, 156, 157, 225
Apiezon pumps 110, 111, 130, 156, 204

Archiv für Electrotechnik 99
argon 211
artificial disintegration 28, 73, 93, 131, 144, 145, 198, 228, 273
Associated Press 253, 254
Aston, F. W. 25, 126
atom 3, 4, 5, 6, 7, 15, 27, 28, 29, 30, 36, 37, 51, 52, 70, 72, 73, 74, 155, 211, 212, 229, 250
atomic bomb 264, 265, 267, 268
Atomic Energy Research Establishment 263
atomic nucleus 39, 54, 165, 176, 216, 241, 257, 261, 262
atomic physics 13, 15, 17, 18, 139
atomic weight 27, 28, 30

Bakelite 130
Baltimore 183, 210
Banks, Joseph 49
Barton, Henry 155
Becker, Herbert 180, 181, 182, 207, 208, 214, 228, 257
Bennett, Ralph 148
Berlin 137, 140, 141, 144, 145, 147, 151, 181, 201, 230, 259, 267
Bernal, J. D. 120
beryllium 28, 41, 161, 166, 181, 182, 208, 210, 211, 213, 214, 219, 228, 243, 257

beta particle 37, 38
beta-rays 51
betatron 65
Bethe, Hans 273
Bishop Foy's School 168
black body radiation 70
Blackett, Patrick 26, 42, 120, 126,
    152–3, 260, 262, 263, 264,
    272
Bletchley Park 272
Bloch, Felix 118
Blumer, Mary 195, 254
Blunt, Anthony 120
Bohr, Niels 69, 72, 73, 77, 83, 84,
    85, 86, 88, 89, 90, 95, 97, 176,
    177, 180, 201, 240, 241, 245,
    257, 262, 272, 273
Born, Max 78, 80, 264, 273
boron 28, 41, 181, 211, 239, 242
Bothe, Walther 140, 141, 180, 181,
    182, 200, 207, 208, 214, 228,
    257, 262, 274
Bowden, Vivian 223
Boyce, Joseph 216, 217, 254
Brasch, Arno 145, 146, 147, 151,
    174, 259, 269, 273
Breit, Gregory 142, 143, 144, 145,
    176, 199, 256
Bristol University 122
Broglie, Louis de 74, 262
Brookhaven National Laboratory
    269, 273
Brown Boveri 146
Browne, Maurice 249, 250
Brussels 262
Bruyne, Norman de 19
Burch, C. R. 'Bill' 110, 111, 156
Burch pump 187, 258, 270

Burgess, Guy 120
Burlington House 49

Calder, Ritchie 270
California Institute of Technology
    (Caltech) 147, 148, 173, 188,
    194, 230, 259, 273
Cambrai 61
Cambridge 7, 9, 10, 15, 17, 18, 19,
    21, 24, 25, 26, 36, 40, 45, 47, 55,
    57, 62, 65, 80, 85, 86, 91, 98,
    101, 103, 112, 114, 123, 127,
    138, 139, 144, 148, 152, 162,
    163, 164, 169, 170, 174, 177,
    184, 189, 190, 199, 201, 203,
    211, 221, 247, 252, 257, 260,
    267, 272
Cambridge Philosophical Society
    117
carbon 27, 41, 157, 211, 243
Carlsberg fellowship 84, 176
Carnegie Institution 50, 143, 145,
    197, 199
cathode 108, 111, 153, 154, 156,
    157, 225
Cavendish Laboratory 7, 10, 11,
    12, 13, 14, 15, 16, 17, 18, 19, 20,
    21, 22, 25, 26, 33, 34, 36, 40, 41,
    42, 43, 44, 45, 47, 52, 56, 57, 58,
    59, 60, 62, 66, 77, 82, 85, 86, 90,
    91, 93, 95, 96, 97, 99, 101, 102,
    103, 105, 107, 110, 111, 112,
    114, 115, 116, 117, 120, 123,
    125, 126, 127, 128, 129, 130,
    131, 132, 136, 137, 139, 141,
    142, 143, 147, 149, 151, 152,
    155, 157, 158, 162, 163, 164,
    167, 169, 173, 176, 177, 182,

183, 185, 201, 202, 203, 205,
208, 210, 212, 216, 226, 230,
231, 235, 236, 238, 244, 245,
247, 249, 250, 252, 256, 257,
258, 259, 260, 261, 262, 263,
265, 270
Cavendish Physical Society 116,
117
CERN 269
Chadwick, James 11, 14, 15, 20,
25, 26, 32, 33, 34, 35, 36, 40, 41,
42, 43, 44, 45, 46, 55, 80, 90,
113, 117, 126, 129, 137, 138,
139, 140, 141, 180, 181, 182,
183, 184, 200, 207, 209, 210,
211, 212, 213, 214, 219, 222,
223, 228, 229, 230, 233, 240,
241, 242, 245, 257, 259, 262,
263, 264, 265, 267, 272
Churchill, Winston 34
cloud chamber 235, 236, 237, 260,
264
Cockcroft, Elizabeth 61, 122, 166,
219, 240, 251, 274
Cockcroft, John Douglas 60–65,
91–4, 96, 97, 99, 100, 102, 103,
104, 106, 107, 108, 109, 110,
111, 112, 117, 122, 123, 124,
125, 126, 127, 128, 129, 131,
136, 139, 142, 144, 148, 149,
151, 153, 155, 158, 159, 160,
161, 163, 165, 166, 167, 171,
172, 173, 175, 177, 180, 184,
185, 186, 187, 189, 192, 196,
199, 201, 202, 203, 205, 207,
209, 210, 215, 216, 217, 218,
219, 223, 224, 228, 229, 230,
231, 232, 234, 235, 238, 240,

242, 243, 245, 246, 247, 248,
250, 251, 252, 253, 256, 257,
259, 261, 262, 263, 265, 267,
268, 274
Cockcroft, Timothy 123, 125
Cockcroft-Walton voltage
multiplier 186, 201, 260, 270
Cold War 265
Coleraine 167, 169, 222
Columbia University 183
Compound Q 204, 215, 259
*Comptes Rendus de l'Académie de
Science* 207, 257
condenser 186, 203, 206, 207
Condon, Edward 87, 90
continuity, principle of 70
Cookstown 167
Copenhagen 73, 77, 83, 85, 86, 88,
89, 90, 92, 93, 176, 178, 201,
245, 273
copper 27
corona 106, 107, 109, 111
Crabtree, Elizabeth *see* Cockcroft,
Elizabeth
Crookes, William 37–8, 49, 136
Crowe, George 22, 158
Crowther, J. G. 178, 179, 202,
244, 245, 252, 253
Curie, Irène 140, 207, 208, 209,
210, 212, 214, 228, 257, 262
Curie, Marie 21, 42, 129, 140, 184,
262
Curie, Pierre 42
cyclotron 150, 191, 192, 194, 195,
196, 217, 222, 254, 259, 265, 273

Dahl, Odd 144, 199
Daily Herald 270

Daily Mail 248
Daily Mirror 249
Dalhousie University 19
Dartmouth University 192
Davy, Humphry 49
Department of Scientific and
    Industrial Research (DSIR)
    162, 163, 167, 218, 220, 221,
    222
DESY 269
Devonshire, Duke of 10
Devonshire family 13
Dirac, Paul 85, 122, 178, 245, 262,
    265
DNA theory 273
Donaghadee 168, 169, 170
Dublin 12, 13, 16, 64, 142, 168,
    189, 220, 261, 263

Eddington, Arthur 252
Edlefsen, Niels 191, 192
Ehrenfest, Paul 85, 97, 98, 144
Einstein, Albert 70, 71, 72, 84, 95,
    137, 234, 252, 273
electrode 108, 206
electromagnetism 14, 52, 192
electron 4, 13, 28, 29, 30, 31, 32,
    33, 36, 37, 51, 52, 58, 59, 64, 69,
    71, 72, 73, 74, 75, 79, 85, 105,
    106, 108, 135, 141, 155, 174,
    179, 180, 208, 260, 269
electroscope 161
Ellis, Charles 137, 140, 263

Feather, Norman 183, 184, 209,
    210, 272
Federal Telegraph Company 196
Fellows of the Royal Society 49, 50

Fermi, Enrico 262
Fermilab 269
First World War 60
fission 267
Fleming, Ambrose 203
fluorine 41, 239, 242
Fok, Vladimir 78
Fowler, Ralph 85, 89, 90, 95, 98,
    117, 122, 176, 178, 243
Free School Lane 10, 14, 112,
    258
Friedrich-Wilhelms University
    145

galvanometer 189
gamma ray 37, 38, 51, 134, 135,
    141, 161, 166, 171, 181, 182,
    184, 188, 208, 211, 212, 214,
    219, 256
Gamow, George 66–7, 72, 75, 76,
    77, 78, 79, 80, 81, 82, 83, 84, 85,
    86, 87, 88, 89, 90, 91, 92, 93, 94,
    96, 97, 98, 99, 117, 139, 144,
    165, 171, 172, 176, 177, 178,
    179, 180, 184, 202, 222, 229,
    232, 260, 262, 272
Geiger, Hans 134, 135, 136, 137,
    140, 200
Geiger counter 135, 141, 180
General Strike 120
Geneva 269
gluons 269
gold 27, 157
Goodlet, Brian 103
Göttingen 73, 74, 77, 78, 79, 80,
    82, 83, 93, 139
gravity 52
Greinacher, Heinrich 131, 219

Guardian see Manchester Guardian
Gurney, Ronald 87, 90

h 70, 72, 74
Hafstad, Larry 144, 199
Hahn, Otto 267
Hamburg 97
Hare and Hounds club 122
Hartree, Douglas 85, 86
Harvard University 150
Harwell 263
Heisenberg, Werner 73, 74, 75,
    85, 86, 144, 245, 262, 265,
    274
helium 28, 29, 31, 33, 73, 211,
    219, 228, 240, 248, 253
Higgs boson 269
Hiroshima 264
Houtermans, Fritz 79, 82, 83
hydrodynamics 13, 17, 57
hydrogen 27, 28, 29, 31, 72, 154,
    208, 210, 211, 228, 240, 253

Institute of Theoretical Physics
    78
Institution of Mechanical
    Engineers 96
Interpretation of the Atom, The 266
Interpretation of Radium, The 266
Irish Free State 17
Irish Independent 250
Irish Times 250
Ising, Gustav 99
Iveagh, Earl of 50

Jazz Band 67, 73,117
Jeans, James 203
Joffe, Abram 76, 77, 83, 202, 262

Johns Hopkins University 142,
    183, 184
Joliot, Frédéric 140, 184, 207, 208,
    209, 210, 212, 214, 228, 257,
    262
Jordan, Pascual 97

Kaiser Wilhelm Institute 140
Kapitza, Peter 56, 63, 77, 91, 119,
    124, 126, 129, 166, 167, 177,
    178, 202, 207, 225, 273
Kapitza Club 117, 118
Kelly Hospital 183, 184, 210
Khariton, Yuli 202
Khvolson, Orest Danilovich 77
Kiel 140
Kirsch, Gerhard 36, 40, 41, 44, 46,
    140

Lange, Fritz 145, 146, 147, 151,
    174, 259, 269, 273
Langevin, Paul 20, 21, 201, 262
Laurence, George 19, 164
Lauritsen, Charles 147, 148, 151,
    173, 174, 187, 217, 239, 255,
    259, 269, 273
Lawrence, Ernest Orlando 149,
    150, 151, 174, 191, 192, 193,
    195, 196, 199, 216, 217, 218,
    222, 230, 239, 242, 253, 254,
    255, 258, 259, 260, 262, 264,
    265, 269, 272, 273
lead 161, 166
Leiden 97
Leipzig 118, 144
Leningrad 76, 83, 202
Leningrad University 73, 76
Lincoln, Fred 113, 128, 129

lithium 28, 41, 161, 181, 211, 219,
  225, 228, 229, 233, 234, 237,
  239, 240, 242, 248, 253, 254,
  255, 256, 257, 265, 267
Liverpool University 182, 263
Livingston, Stanley 192, 193, 196,
  216, 222, 242, 255, 256, 269, 273
Lodge, Oliver 121, 203

magnesium 27, 41, 181
magnetism 53, 56, 68, 193
Manchester 3, 21, 22, 24, 25, 37,
  59, 63, 69, 95, 124, 130, 172,
  229, 240
Manchester College of
  Technology 62
Manchester Guardian 202, 244
Manchester University 60, 137
Manhattan Project 272
Marconi, Guglielmo 201
Marsden, Ernest 3, 4, 5, 6, 7, 37,
  43
Martin, Leslie 59
Massachusetts Institute of
  Technology (MIT) 197, 198
Massey, H. S. W. 'Harrie' 240
Maxwell, James Clerk 13, 52, 117,
  201, 203, 205, 222
McKerrow, George 114, 173, 240
Meitner, Lise 140, 141, 181, 200,
  262
mercury 27, 110
mercury diffusion pump 110
Metropolitan-Vickers 59, 62, 63,
  91, 103, 104, 110, 114, 123, 124,
  162, 172, 173, 175, 202, 203,
  204, 206, 207, 218, 222, 240,
  258

Meyer, Stefan 40–41, 46
Millikan, Robert 193, 201, 257
Mr Tompkins Explores the Atom 273
molybdenum 109
Monte Generoso 146
Montreal 21
Morse key 43
Mott, Nevill 85, 86, 89, 98, 240,
  262, 263, 272
muons 269

Nagasaki 268
National Academy of Sciences 191
National Physical Laboratory 222
Nature 40, 87, 88, 89, 90, 91, 214,
  217, 230, 237, 239, 241, 242,
  246, 247
Naturwissenschaften, Die 180
neon 6
Nernst, Walther 146
neutrinos 269
neutron 32, 47, 179, 181, 182, 213,
  214, 228, 230, 242, 245, 257,
  260, 267, 269
New Statesman 121
New York 183
New York Times 174, 253, 254
Newton, Isaac 49, 52, 68, 272
Newton-John, Olivia 273
Nichols, Robert 249, 250
Niedergesass, Felix 107
nitrogen 39, 41, 60, 157, 211, 212,
  243
Nobel prize 3, 6, 25, 56, 83, 122,
  190, 261, 262, 263, 272, 273
nucleus 5, 7, 25, 28, 30, 31, 33, 34,
  36, 37, 39, 40, 41, 47, 53, 54, 58,
  69, 72, 80, 81, 82, 87, 88, 90, 92,

93, 98, 139, 142, 145, 149, 151,
154, 155, 161, 165, 166, 172,
176, 178, 179, 195, 199, 208,
210, 213, 214, 222, 228, 229,
232, 233, 234, 236, 241, 248,
250, 257, 259, 267, 268, 269
Nursery, The 18, 57, 63, 136, 216,
226
Nutt, Horace 212

O'Casey, Sean 26
Occhialini, Giuseppe 260
Odessa 66, 67, 86
Oliphant, Mark 12, 18, 24, 126,
259, 263, 264, 272
One, Two, Three . . . Infinity 273
Order of Merit 7
Oxford University Press 178, 179,
180, 244
oxygen 27, 28, 29, 211

Painlevé, Paul 118
Paris 140, 201, 207
particle physics 269, 270
Pasadena 147, 148, 151, 173, 217
Pauli, Wolfgang 85, 86, 144, 262,
264
Peierls, Rudolf 118
Petrograd 67
Pettersson, Hans 36, 40, 41, 44, 45,
46, 140
Philosophical Magazine 80
phosphorus 27, 41
photoelectric effect 71
photon 208, 212
Physical Review 148, 174, 195, 216,
222, 260
Physico-Technical Institute 76

pions 269
Planck, Max 70, 71, 72, 201
plasticine 188, 204
platinum 28
Plough and the Stars, The 26
plutonium 268
polonium 140, 141, 180, 181, 182,
184, 207, 208, 210, 214, 228,
232, 257
positron 259, 260, 269
potassium 27
Prince Edward Island 19
Princeton University 87, 142, 197
*Proceedings of the Cambridge
Philosophical Society* 119
Proceedings of the Royal Society 49,
55, 165, 172, 218
proton 28, 29, 30, 31, 32, 33, 39,
40, 92, 93, 94, 96, 97, 99, 104,
111, 132, 143, 154, 156, 157,
159, 161, 172, 179, 181, 184,
193, 194, 199, 206, 208, 211,
213, 214, 216, 217, 218, 219,
223, 225, 228, 232, 233, 234,
239, 251, 269, 270

quantum theory 70, 71, 72, 75, 80,
81, 82, 83, 86, 95, 96, 144, 176,
179, 189, 232, 264
quarks 269

radioactivity 4, 15, 21, 30, 37, 38,
39, 42, 51, 53, 80, 136, 137, 141,
157, 159, 161, 179, 181, 183, 184,
207, 211, 213, 228, 236, 242
radium 30, 37, 38, 42, 46, 51, 52,
53, 54, 130, 140, 141, 158, 179,
208, 228, 232, 256

Radium Institute 140, 208
Rayleigh, Lord John William
    Strutt 13, 117
rectifier 105, 106, 108, 111, 130,
    153, 159, 187, 188, 203, 206,
    207, 225, 256, 270
Red Army 67
Reichsanstalt 140, 141
relativity, theory of 86, 234
Reynolds's Illustrated News 246, 247,
    249, 253
Rhodes Scholar 196
Rockefeller fellowship 176, 178
Rome 201
Royal Commission 16
Royal Danish Academy of
    Sciences 84
Royal Field Artillery 60
Royal Institution 124
Royal Society London 4, 7, 49,
    51, 56, 57, 58, 60, 92, 98, 102,
    120, 124, 135, 143, 165, 167,
    173, 176, 196, 218, 220, 241,
    243, 246
Rutherford, Eileen 85, 89
Rutherford, Ernest 2, 3, 4, 5, 6, 7,
    11, 12, 13, 14, 15, 16, 17, 18, 20,
    21, 22, 23, 24, 25, 26, 30, 31, 32,
    33, 34, 35, 36, 37, 39, 40, 41, 42,
    43, 44, 45, 47, 49, 50, 51, 52, 53,
    55, 56, 57, 58, 59, 60, 62, 63, 64,
    65, 67, 69, 72, 73, 80, 81, 82, 85,
    89, 90, 92, 93, 94, 95, 96, 98, 99,
    100, 102, 103, 104, 110, 112,
    113, 115, 116, 117, 118, 119,
    120, 126, 127, 128, 129, 132,
    133, 134, 135, 136, 137, 138,
    139, 140, 141, 143, 144, 145,
    158, 160, 162, 163, 167, 173,
    174, 176, 177, 180, 181, 183,
    185, 191, 197, 200, 201, 203,
    207, 209, 213, 214, 219, 221,
    222, 223, 224, 227, 228, 229,
    230, 235, 236, 238, 239, 240,
    241, 242, 243, 244, 245, 246,
    248, 252, 253, 257, 258, 259,
    261, 262, 263, 264, 265, 267,
    269, 270, 271, 272, 273
Rutherford, Mary 23, 112, 116,
    120, 177

Sargent, Bern 136
Schenkel, M. 186
Schrödinger, Erwin 74, 75, 81, 85,
    88, 144, 262, 265, 273
scintillation 37, 38, 39, 41, 42, 43,
    45, 47, 132, 134, 136, 141, 183,
    184, 225, 226, 227, 231, 235,
    237, 242, 256, 270
Scott, C. P. 244
Searle, G. F. C. 18
silicon 28, 41
Snow, C. P. 235, 263
Soddy, Frederick 266
sodium 27, 41
solar system model 5
Solvay conference 262, 272
Southern California Edison 147
Soviet nuclear weapons project
    273
spinthariscope 38
Stalin, Joseph 202, 273
Stebbings, Joyce 126
Stern, Otto 150
Strassmann, Fritz 267
sulphur 6, 27

Sunday Express 247
Swirles, Bertha 178

Terrestrial Magnetism,
    Department of 143
Tesla transformer 50, 59, 106, 108,
    143, 145, 151, 153, 197, 199
thermionic valves 132
Thomson, J. J. 13, 21, 49, 58, 80,
    113, 114, 118, 119, 203
Todmorden 60, 61, 122, 220, 251,
    252
transformer 51, 52, 59, 102, 103,
    104, 130, 154, 155, 159, 161,
    166, 171, 175, 184, 186, 192,
    206
transubstantiation 66
Trinity College Cambridge 11, 23,
    214
Trinity College Dublin 12, 18,
    73, 162, 168, 220, 221, 263,
    264
Tuve, Merle 142, 143, 144, 145,
    148, 149, 151, 174, 197, 198,
    199, 200, 216, 217, 230, 239,
    242, 254, 255, 256, 259, 269,
    273

uncertainty principle 75
University College of North
    Wales 131
University of California at
    Berkeley 149, 174, 196, 199,
    217, 230, 254, 256, 273
University of Minnesota 142
uranium 28, 29, 30, 179, 264, 267,
    268
Urban, Kurt 145, 146

vacuum pump 109, 113, 138, 192
Van de Graaff, Robert 197, 198,
    199
Van de Graaff generator 198, 200,
    216, 217, 255, 260
Vienna 36, 40, 41, 44, 45, 46, 140,
    144
voltage 50–52, 54–5, 59, 60, 92,
    97, 102, 103, 106, 107, 108, 145,
    146, 148, 153, 154, 156, 157,
    158, 159, 160, 161, 172, 174,
    175, 184, 185, 186, 188, 196,
    198, 215, 216, 217, 225, 227,
    229, 230, 231, 237, 241

Wall Street Crash 114
Walton, Ernest Thomas Sinton 9,
    10, 11, 13, 16, 17, 18, 19, 24, 57,
    58, 59, 60, 64, 73, 96, 97, 99,
    100, 102, 103, 106, 107, 108,
    109, 110, 111, 112, 117, 118,
    119, 121–2, 125, 127, 128, 129,
    131, 136, 139, 142, 144, 148,
    149, 150, 151, 153, 155, 158,
    160, 161, 162, 163, 164, 166–71,
    172, 173, 175, 177, 184, 185,
    186, 187, 189, 190, 191, 192,
    193, 195, 196, 199, 201, 203,
    205, 207, 209, 210, 215, 217,
    218, 220, 221, 222, 223, 224,
    226, 227, 228, 229, 230, 231,
    232, 234, 240, 242, 243, 244,
    245, 246, 247, 248, 250, 251,
    252, 253, 256, 257, 258, 259,
    261, 262, 264, 265, 267, 268,
    274
Walton, Jim 251
Walton, John 167–8

Walton, Winifred (Freda) 168, 169, 170, 171, 185, 189, 190, 191, 195, 205, 218, 220, 221, 222, 238, 240, 250, 251, 252, 263, 274

Ward, F. A. B. 132, 133, 135, 136

Washington, DC 50, 144, 145, 174, 196, 197, 199, 217, 230, 256

Waterford 168, 169, 189, 250

wave function 88

wave theory 71, 72, 96

Webster, Hugh 181, 208

Wells, H. G. 265, 266, 269

Wesleyan Society 121

Westminster Abbey 272

Widerøë, Rolf 99, 149, 150

Wigner, Eugene 79, 82, 144

Wilson, C. T. R. 25, 190, 235

Wilson, Winifred *see* Walton, Winifred

Wings Over Europe 249, 266

Wood, Alec 121

World Made Free, The 266

Wren, Christopher 49

Wynn-William, Eryl 131, 132, 133, 134, 135, 136, 210, 227, 231, 239, 270, 272

X-ray 50, 51, 103, 107, 147, 148, 149, 151, 159, 174, 217, 259

Yale University 149

Zeeman, Pieter 201

Zeitschrift für Physik 82, 83, 84, 87, 88, 89, 91

zinc 6

zinc sulphide 38, 132, 181, 225, 231, 235, 270

Zurich 144